This World

Written by
Ron Humphrey

Dedicated to my brother, Ed and sisters, Jo and Chris.

Published by Sweetfields Publishing

First published 2019 Sweetfields Publishing
Wingham, N.S.W. Australia 2429.

email:- maureenlarter@gmail.com

Cover by Christine K Dunkan, Moonlit Magic, including free images from the Internet. email: christinkyrin@gmail.com

ISBN: 978-0-6484695-8-2

About the author.

Ron Humphrey was born in United States in the late 1950s. After working as a construction worker, he moved to Australia in the 1980 and took up life in New South Wales. This book has been written over the last couple of years, and represents the fulfilment of the forecasts of the scientists and weathermen. It came about after a unsettling dream, and Ron was encouraged to write by his siblings.

Chapter One
Surviving in the New Climate

Whether global warming was man-made or a natural event didn't much matter when it came to rising sea levels. The year was 2091, the world had changed, and humans had adapted to a changing world.

Most waterfront properties from the early and mid-twentieth century were now underwater, and the landscape had changed all around the world. Humans can be stubborn, and many tried to save their ocean-front homes or lakeside cottages, but most were lost in one tsunami or another and to the tides. Japan had been devastated after another tsunami had caused another nuclear meltdown.

Oddly, the islands of Bali, the Philippines, and Thailand had been rising with the sea. It seemed the rising water was pushing those islands higher due to a root system holding together and growing rapidly. This process produced rapid plant growth on top of the soil, too, which provided bumper crops in some parts, though they were being devastated by Japan's nuclear fallout in other parts. Apparently, when some winds blew over the mountains, only people on one side were affected. As it blew over the other side, it stayed high enough in the atmosphere to not affect those living there, and those people survived. Much of Indonesia had been destroyed in a tsunami, much like the one they'd had in 2003. In the Americas, the water completely covered Florida and Louisiana, and the Panama Canal was now ten nautical miles wide.

Ron, a sixty-year-old retired accountant and construction worker before the devastations had stayed in his house in the small town of Wingham, Australia, which only became a waterfront property after the rising sea levels. The creek was one hundred meters away when he'd bought the place. That had been thirty years ago. Now the house was five meters from the water, which had also claimed some of the lower backyard. The water came up to the property line and fence at the side of the house and covered what used to be the veggie garden at the back of the yard.

There were five more houses up the street, as a hill rose to the corner, and then seven more houses around the corner. Across the road, there was the old produce store with the grain silos that the town had come to rely on. What used to be a thriving frame and truss factory had been abandoned and turned into a large chicken coop. The locals in the street all shared care of the free-range chickens and the eggs. The road that used to run to the other side of town now ended at his house. In the other direction, the road had only needed to be changed slightly to find higher ground and get over the railway tracks. It was a peninsula on the edge of town, and the entire town had become an island that wouldn't have survived at all until they had built a new and higher bridge. There were a lot of people living in the center of town these days, probably a lot more than used to before the sea levels rose. People living on higher ground outside of the town used small boats to visit or do their shopping.

The town's attractiveness was due to its connection to the roads and railway station. Many other towns had become isolated by the water, and for them, being able to get food and supplies into town had become too difficult, so they'd become ghost towns. Most people still lived in the cities, although cities on the coast had been decimated to the extent that most high-rise buildings were vacant and surrounded by water. The rail station was a big asset to the town because more than half the stations in the state had closed since the last time the water had risen. You could get a train to Penrith and then Canberra. Canberra had survived almost unchanged; there was some loss of land to the water, but only a few roads had needed to be changed, and only a few buildings had been lost. Ron's son, daughter-in-law, and grandchildren lived in Canberra.

There was no longer a rail connection between NSW and Queensland. The NSW rail from Sydney stopped at Taree, and the Queensland rail only went from Brisbane to the west. There were still trucks able to get goods interstate, but there were detours, and it took a lot more time than it had previously. There had been many road closures, and the new roads weren't always tarred. Some areas just had too much water, and many towns all around Australia, and also around the world, were abandoned.

The small town's grain silos were also a big asset and had kept the locals fed when the roads were closed for months at a time. People were able to live on higher ground outside of town, but some needed a boat to get into town for supplies or to socialize. Small drones had been used to fly supplies into some places, but they couldn't carry much weight and were often shot down en route and the supplies stolen. They would still use drones in the cities, but they had given up trying to service the state with them.

Ron was reasonably comfortable where he was living, notwithstanding his physical condition and chronic pain. He had broken his leg, pelvis, and ribs, crushed his spine, punctured his lung, and taken a few hits to the head after falling over eighty feet while working as a rigger at the power station twenty years before. He was lucky he hadn't been a free fall; he'd managed to bounce off a few steel beams and through a swinging scaffold, where he'd been able to grab the steel rope and slow himself down considerably. He'd spent a lot of time in the hospital and more in rehabilitation.

After he'd left the hospital, he'd been able to manage most jobs as long as the pace was slow and steady, but he wasn't able to do the things he used too. Things like running or jumping were almost impossible, and he needed to lie down for ten or twenty minutes a few times a day. He couldn't lift the weight he used to, but that happens with age anyway. When he would stay seated for more than an hour, his back would become stiff, and he could hardly straighten it up. The slipped disk and arthritis from all the broken bones were the most constant pain, but it was when his ruptured bowel and diaphragm had caused his lung to collapse that was most painful. Thankfully, that didn't happen too often, because, when it did, it felt like he was going to die, and even though he could breathe again after a few minutes, he would be sore for days. Every day, he was certain to be sore by the time he went to bed, but he felt lucky he was still surviving.

He had a two-thousand-liter rainwater tank with a pipe connected to the outside toilet, so when there wasn't town water, he could still flush the toilet. There hadn't been any town water for almost two weeks. The town folk would rather the tap be dry than let the contaminated stuff through. He would only drink and cook with the tank water, but others used it to wash dishes or shower. He also had six solar panels and a stand-alone battery system that powered a refrigerator and lights, and he used it to recharge other batteries for his laptop and various appliances. There was still a power grid, but it sometimes shut down for weeks and didn't have the lines to reach everywhere. Solar was his only power, and because heating up an oven used so much of it, he seldom cooked indoors. He made his own brick fireplace in the backyard with a steel grill plate and cooked his meals over the wood fire. Of course, he couldn't cook out there if it was raining too hard, but he seldom cooked anyway. He preferred not to cook most of his food, which was usually cereals, fruits, and vegetables.

He had an orange tree, a few grapevines, and blueberries and strawberries for fruit. He grew lettuce in boxes and kept peas and green beans growing in constant cycles. He had tomatoes in pots and a little hemp patch. He thanked goodness they'd legalized hemp, because it was the only pain relief he had. When food was scarce, he ate hemp, and he'd decided he liked it. He added hemp leaf to rice and threw a few flower heads in his salads. A lot of people grew it, and some commercial growers found all kinds of uses for it. They made hemp rope and clothes like in the old days and also made fiberboards that could be used for building material. It was the easiest crop to grow in the new climate, so almost everyone had a patch, and if they didn't use it themselves, they could sell or trade it for other goods.

One of the neighbors on the little peninsula kept bees and provided honey. They would all share veggies and fruit, and he would share honey. The neighbors who looked after the hens for eggs sometimes grew chooks as well, and when there was more than needed, they would sell them. They'd set up a table near the road in front of the chook coop and put up a sign. They would sell the honey there, too, and if anyone else had veggies they wanted to sell, they would bring them along as well. They used to do a reasonable business with people coming to the produce store next door for supplies, who would then pick up a few things from their table. There were six of them sharing the chooks, and when they sold the eggs, chooks, or manure, all the money went into a can. Every Sunday morning, they would have a meeting at the table and see how much they had sold and if they needed to buy any feed or supplies. If there was enough money to cover expenses, they thought that was good, and if there was extra, they would split it up.

Ron was at home watching the evening news, and it occurred to him that when things are going so well is when they seem to get worse soon enough. The breaking story was Mount Etna had erupted, as it does, but this time in a big way, triggering a huge earthquake, and in the past hour, there had been several aftershocks. There was a picture from the sky, and the eruption looked to have wiped out every living thing and knocked down every building for at least a hundred miles around it. Just as the story ended, the newscaster motioned that he was hearing something from his earpiece and to hold on. Breaking news was that Japan's volcanoes had all erupted at the same time. There had been some huge news stories regarding wild weather, earthquakes, tsunamis, and the rising water in the past, but this seemed even more serious. There was a warning to be prepared for a tidal wave and an upward surge in the water table. That was actually a pretty common thing for them to say over the past few decades, and it hadn't always been the disaster they'd predicted, so he took the news in stride.

He ate his supper of two cold hard-boiled eggs with chopped lettuce, onion, hemp head, and a beautiful fresh tomato. He played the piano for thirty or forty minutes and then turned the news back on. Aftershocks were slowing but continuing, and not just around what used to be Mount Etna but throughout all of the Italian and French Alps. Aftershocks were also hammering Japan. Ron thought one good thing was almost all of the Japanese had already left the Island because of the previous nuclear tragedy, so there wouldn't be much loss of life there. He lay down on his mat on the floor in front of the TV and started his physiotherapy exercises while the news kept showing pictures from a helicopter that was so far away from the volcano all he could really see was a huge cloud. He guessed to himself the earthquake must have knocked out their power, towers, or something, or there would have been pictures from the locals there.

He finished his exercises, closed his eyes, and tried to relax everything. Then there was more breaking news: Mount St. Helens in the U.S. had erupted. He thought to himself that was impossible. It had been destroyed long ago or been dormant or something, hadn't it? The news was just breaking, and there were no pictures, but the reports said it was huge, and another earthquake followed.

Then the news flashed again, and they had pictures of a US anchorman on the national news being shaken all over the place, with the ceiling, lights, and cameras falling and breaking behind him. The picture went blank and then, about thirty seconds later, came back on again. The newscaster and a few other people were standing up, the lights were dim, and there was dust all over them and the desk. They all looked to be in shock, and then there was an aftershock. The camera kept rolling while the room shook, and the anchorman and crew must have ducked under a desk, out of the picture. When the shaking stopped, they all got up with these great, big, wide-open eyes and looks of shock on their faces. There was no sound, and the camera must have been on automatic record. They were just standing there, looking around and at each other with those looks of shock. Ron thought to himself it was hilarious, but then he realized he must be becoming rather insensitive, and he nearly cried.

About a minute later, the news flashed back to the Australian news desk, and they had satellite pictures of Japan. The pictures were just a lot of debris scattered in the water. It looked as if Japan was no more. They said the eruption and ensuing earthquakes had shaken what was left of the mountains until they had crumbled and fallen into the sea. They said lava was still erupting into the sea, and the warming was causing the water to boil. It was reported there were around five thousand Japanese still living there in the mountains and all were presumed dead.

Most Japanese had moved to China, but there were also many in Australia. Ron again almost cried, thinking of those he knew, and he wondered how they were taking the news.

They also had more pictures of Mount Etna, or where Etna had been, and footage showing landslides in the French and Italian Alps. They were still having aftershocks around every hour. The pictures emerged of the American Northwest, and they were also having earthquakes in Washington State, Oregon, and what was left of the Sierra Nevada in California. Most of California had been lost to the sea, but the Sierra Nevada had been considered relatively safe and had become one of the most populated places in the US. They were showing ash clouds from all the volcanoes and said all the clouds had blown north. The ice cap had melted away considerably years ago but had been gaining in area the past several years. The anchor spoke to an environmentalist, who predicted this could have an adverse effect on the remaining ice cap and they could expect the sea levels to keep rising for another long while.

Then there was another breaking news bulletin: volcanoes had erupted in Hawaii. No pictures were available yet, but it was reported it would likely be the end of Hawaii as most of what hadn't been lost to the sea was uninhabitable because of the volcanoes. Some mountains on all the Islands had survived and were heavily populated. Ron thought they would all have to wait and see, and he hoped maybe some might survive. He was becoming indifferent to life and death, but every once in a while, just for a minute or two, these types of horrible stories would make him cry. Then he would say to himself, *That's enough* and *We all have to carry on*.

Chapter Two
There's Always Another Morning

The next morning, Ron woke around 8a.m.It looked to be a sunny day. He walked to the toilet for the morning's first order of business and then made himself a cup of coffee. He had lived alone since losing his wife fifteen years earlier, and the morning always started with a cup of coffee and the time to enjoy it. That was also when he thought ahead about the day's necessary chores and how he would be spending it. After his coffee, he picked up his compost container he kept next to the sink, a two-liter plastic bowl with a lid, and noticed it was rather full. It contained used coffee grounds, eggshells, and the like. There was also a separate ten-liter bucket containing the unusable outer lettuce leaves and shells from the peas and broad beans after he had shucked them.

He walked out the back door and put the large bucket on the picnic table. Then he walked the few steps to the compost bin, opened it, dropped the contents of the plastic bowl into it, and put the lid back on all in one motion. He had emptied that container nearly every morning for the past twenty years. He walked over to the other bucket, picked it up, and carried it across the road to feed the chooks in the factory they had converted into a chicken coop. He opened the coop door, and they all pushed out in a hurry, jumping up to the bucket and trying to peck at it before he could empty it. He spread out the contents, and then the rest of the chooks chased the feed with vigor. There were at least a hundred layers, and they were producing two hundred eggs or more each morning.

Ron then took the bucket over to one of the large, lidded feed barrels they kept next to the coop, opened it, and started scooping chook feed pellets until the bucket was full. He then dropped the pellets in a line for about ten meters. He went back and loaded another bucket and walked over to where the road met the water at the front of the block. About ten chooks followed him, and he spread the contents among the three-foot-tall patch of weeds. The chooks would knock those weeds down, which would tidy up the area at the same time. The chooks roamed free every day but tended to stray to the back of the block. There wasn't much growing in the yard but still plenty of food for them to find and digging to do at the back.

When the group decided they wanted chicks, they incubated the eggs in the shed at the back of the block. They would hatch a few hundred at a time and had a fenced yard around the shed where the chicks spent the night. The chicks stayed together at the back of the block and stayed close to their yard. At six weeks old, they would be sold for ten dollars each, and out of the original two hundred, there would be around one hundred that survived to that age. During the day, the chooks went where they wanted and cleaned up the weeds at the back of the block right up to the water's edge. The layers would go to the back of the block but not as far, and they always stayed together. Evenings, they would head to the coop expecting another feed and get locked in for the night. They only hatched chicks four times a year. Otherwise, it was too much work, and that also gave the weeds and grass time to recover.

Sue, a neighbor and partner in raising the chooks, arrived. She had a bucket of green waste, too, and was leaning to one side as she walked with the weight in one hand. She set the bucket down, knocked it over, and rolled it over with her foot, spilling wilted lettuce leaves and carrot tops in front of the ten or twelve chooks that were racing towards her, looking for a green treat.

Ron said, "I just gave them two buckets of pellets and a bucket of green, so they're eating well today."

"The eggs have been getting bigger," replied Sue," and the yolks are deep yellow, so it seems to be worth the trouble."

Sue was a married woman around forty years old, short and petite with long, straight, shiny black hair and dark eyes. If she hadn't been married, Ron would have been obsessed with her, but he would never covet a married woman. He did like her company, though, and thought if anything ever happened to her husband, he might have a chance to be more than her friend. He knew that even if she wasn't married, she was out of his league, though, and would probably want someone else. He was twenty years older and not good looking or rich. Still, they liked each other's company and were friends and workmates when it came to the chooks.

Her husband, Charlie, was a good guy who worked at the abattoir. He helped out with chooks on weekends and had been very handy when it was time to build the fences and make the coop fox-proof. Ron wouldn't want to come between them even if he could.

Ron and Sue collected the eggs and then placed them neatly in the containers. When they sold eggs at the table, they asked people to use their own containers or, if they needed to take one of theirs, to bring it back empty when they wanted more. They weren't commercial growers, and the cost of those containers added up if you had to pay for them. Whenever any of them caring for the chooks wanted one or two dozen eggs, they took them. The money from the eggs sold was put into a tin and sorted out at the Sunday meeting.

Hazel, a rather large eighty-eight-year-old woman, another neighbor caring for the chooks, spent a lot of her time at the table. She probably thought it was the least she could do considering she wasn't any help with most of the chores. She waddled over to the table with her knitting in one arm, holding it close to her chest, and waved with the other arm. Ron and Sue waved and carried the eggs over to the table.

Hazel asked, "Can you do me a favor? I have a wheelbarrow full of pumpkins; will you go bring it over for me?"

Ron said, "Sure."

Sue walked with him across the road and said, "Did you see the news this morning?"

"No," replied Ron. "I didn't want to start crying first thing in the morning." And then he gave a little laugh and added, "It was awful news last night. What was the news this morning?"

She told him the report had said volcanic dust and heated water from the lava were being pushed into the atmosphere and warming ocean currents were moving towards the North Pole. There was another tidal wave heading toward North America, and they expected that the melting ice would raise the sea level more than a meter over a few days and that it would be another catastrophe. Some ecologist they'd interviewed had said they couldn't really know how high it would rise and that it could be more.

Ron said, "You would never know the world is in so much strife just standing here looking out at a beautiful sunny day. I suppose this won't last long either, then. These kinds of events seem to reach us sooner or later. I'll have to get home and turn on the news as soon as we unload these pumpkins."

After Ron wheeled Hazel's empty barrow back to her front yard, he went home and turned on the TV news. There were pictures of water flowing and carrying houses and cars with it in North America. The tidal wave had pushed debris right up the St. Lawrence River, and the melting ice from the north had caused massive flooding. It wasn't a heavily populated part of the States after a recent nuclear disaster, and it would be even less populated now. From the satellite photos, the St. Lawrence and the Great Lakes all looked like one sea inlet from the Atlantic.

They said the Mississippi was now breaking its banks on both sides of the river and the floodwaters from the Great Lakes were moving in a southwest direction. Ron thought, *Here we go again with more water taking over the land and all the news stories of heartbreak and the human will to survive.* He thought they overstated the bit about humans' will to survive. If the global warming and sea levels hadn't have caused all the death and chaos, the human race would have probably wiped itself out with the nuclear wars. North Korea had become uninhabitable, as well as parts of Europe and North America. Iran had been almost completely destroyed, and the nuclear clouds had blown all over the Middle East. Even where humans could survive, they were being born with all kinds of defects. It was only when the sea levels flooded everyone's military bases in the Pacific that they stopped dropping the bombs.

He thought that if the rising sea levels hadn't given them all something more urgent to do, they would probably still be dropping nuclear bombs or be dead, so in a way, global warming had saved the human race. He thought it was a positive way of looking at it, but try telling that to someone who had lost their family, home, town, or country. He felt lucky to have survived and still have his house, his son, and his family doing well. The world was in awful strife, but in his little part of the world, he was still healthy, happy, and secure. At least, he was happy as long as he didn't turn on the news or think about what was happening on the other side of the world.

Then the news showed new pictures of what used to be Mount Etna and said tremors have become less frequent but hadn't stopped. Aftershocks were still rocking the French and Italian Alps and Rocky Mountains in the U.S. They showed pictures of Hawaii, and it was mostly debris floating in the water. It looked like half of a mountain had fallen away so that one side was a cliff and the other side untouched. He thought, *I get the picture, more death and despair with a few lucky souls*, and turned off the TV. He looked out the window, and it was still a beautiful, sunny day. *Winter on the Mid North Coast is sensational*, he thought, *and you would never know just looking outside today that the world is in such an awful state.*

He decided he should enjoy the day despite what was happening. He did the rounds around the veggie patch, searching for and removing snails and just checking the health of the plants. He went to the water tank, filled a watering can, and watered the lettuce so it would perk up and stand at attention. He would come back later and grab some leaves for lunch and dinner. If he didn't water them, then they would be shriveled and limp, still edible but not as good. He went back inside and came out with a large colander and filled it with broad beans. He thought that because he still had plenty in the refrigerator, maybe he would do some canning or give them away. He took the broad beans back into the house and dropped them into the crisper. It was full to the top. *No worries*, he thought. They would stay good for a month or so, and he could eat them every day until then. The plants would still be producing for at least that long, so he would sell or give away the rest.

Steve and Deb next door grew a lot of tomatoes, and the young kids that lived in the next house over had an orange tree and also did a lot of fishing. Now and then, they would have a big catch and give some to the neighbors. Whenever they did, neighbors were so grateful they would share whatever they had straight away. Not as much as a trade but just to show gratitude. Professional fishermen and fish farmers couldn't provide enough to keep city restaurants and grocers in supply, so the only fish country folk usually ate were the ones they caught.

Ron thought maybe he would set his yabby trap. Yabbies had thrived after the rising sea levels, probably because of all the dead animals that had ended up in the water. Farm animals, feral animals, wild animals, and even humans had perished in great numbers. There were plenty of fish kills too with all the flooding and mud in the water. The scavengers like yabbies that survived sure had a lot of food.

Weeks before ,at the edge of the water, in the mud near his backyard ,Ron had found a pallet of boxes containing cans of cat food. Apparently, it had dropped into the water during a tornado, and the flood water had moved it to his backyard. He didn't have a cat, but he could use the cans in his crab trap. The crap trap worked just as well for yabbies.

He walked to the garden shed, picked up the trap, and set it on the picnic table. He went back inside and grabbed six cans of cat food, some steel wire, a pair of pliers, and a screwdriver. He walked back outside and put the wire and cans on the picnic table. Then he picked up one of the cans and stabbed it with the screwdriver. He made several tiny holes and one big hole on the top and bottom. Then he pushed the end of the wire through into one hole and out the other. He used the pliers to cut the wire so there was about a foot on each side. Then he used the wire to tie it at the bottom of his steel crab trap. The trap was about a meter long, about half as wide, and a foot deep. He tied one can in the middle and one can in each of the four corners. He dropped the trap into the creek and tethered it to his backyard fence.

Later that day, he had emptied the trap twice and had almost a hundred yabbies. He started a fire in his outdoor fireplace and went into the house to find a boiler to cook them. He kept around twenty yabbies aside to keep in the fridge for a meal another day. There was more than enough for the neighbors that night, and twenty yabbies would make a few good meals for one. He walked across the road over to the table where Hazel was sitting and let her know he had yabbies and would cook some up later if anyone wanted to come over and bring a plate of something. Then he turned around and went home.

He returned to the fire with a boiler full of water and put it on top of the steel plate on top of the fire. He fed the fire a little more wood, and ten minutes later, the water was boiling. He poured the yabbies from his bucket into the boiling water. Some people thought that was cruel, but he figured the yabbies died so quickly it was more humane than most deaths. Steve and Deb walked into his backyard.

Steve said, "Hazel tells us you're cooking up some yabbies tonight, so we brought a bottle of wine and potato salad."

"Sounds good," Ron replied."Hazel, Sue, and Charlie should be over soon."

The yabbies boiled for a few minutes, and Ron took the pot off the fire. Charlie, Sue, and Hazel, who was struggling to keep up behind them with her walker, came into the backyard. Charlie was carrying a bucket of unhusked sweet corn. He said it was from his veggie garden. They stood around the fire on an absolutely still, clear night, talking and laughing, and all agreed it had been a beautiful day and evening so far.

Later, they sat around the picnic table eating yabbies, sweet corn, broad beans, tomatoes, and potato salad.

Charlie said, "Did we all see the news tonight?"

"No," said Ron."I was out here doing this; what is it now?"

"More volcanoes and earthquakes. This time, it sounds worse than usual, and that's saying something." Charlie shook his head. "They said the Andes would practically disintegrate and crumble into the ocean and the land will displace water and the sea levels will rise in Australia at least another three meters because of it. They said, when it happens, it will be so quick sea levels will raise that much over one night."

"Shit."

"They said the floodwaters in the U.S. are pouring into the Grand Canyon from both sides and is causing a huge current and that is feeding the current to the south. There are also more tsunami warnings from the earthquakes and eruptions in Japan and New Zealand, and those are causing more water displacements with their mountains crumbling into the sea."

"It sounds pretty scary this time."

"No shit. It's expected we will see some changes on our shores before morning, and there's no telling how fast or exactly what happens after that."

"I can guess what it means for us: another three meters of water, and it will be up to my roof and will cover almost all of Wingham. I guess we have to go somewhere else and pretty soon."

Steve said, "They said Canberra and Uluru are still safe places to go. We could go to Uluru but will need a boat, and it's a long way to row. A motorboat will be useless when you can't buy any fuel along the way or carry as much fuel as you would need."

Steve told Ron how he had been thinking about it and how there were also supplies like water, food, and shelter they would need to take with them, and with so many other people with the same idea, they would probably be shoulder to shoulder over the entire rock.

"Could we drive to Canberra?" Ron asked.

"Maybe, but there is only one road, and everyone else will have that same idea, too. Even if the road doesn't flood, we would still be at the back of a very long line."

Ron looked at his little tin boat. He knew Steve and Deb had kayaks, and Charlie and Sue had an ocean-worthy fiberglass boat.

He said, "We could take boats and drive as far as we can. When we can't drive any further, we can get in the boats and go the rest of the way."

For almost a full minute, they all looked back and forth at each other, thinking.

Charlie said, "We should all stick together; my boat will be useless because we won't be able to carry enough fuel. I've got a little outboard motor that we could put on your tinny, and that would get us a lot further."

"Should we ask the rest of the neighbors what they're going to do?" asked Ron.

"How about we ask them if they have boats first and then if they want to come along. Your little tinny and two kayaks won't leave much room for anyone else, and we need to take gear. I'd say only ask them to come with us if they have a boat and it doesn't rely on fuel."

Steve said, "There are at least four long rowboats with oars underwater at Taree's old rowing club. I don't think anyone would have thought to dive underwater to retrieve them. They're probably still under the veranda behind a steel cage, and if we could get them, we could take the whole neighborhood."

"I have flippers, a mask, and a weight belt if we want to have a look," said Ron.

Taree had been flooded for some years, and the riverbank had slowly widened all the way to the street next to the hospital. The tops of some of the buildings were still visible, and the buildings were intact. Their floors above the waterline had become a home for squatters. People had taken residence in the top floors much like those in the big cities had done in skyscrapers.

They needed a boat to leave the building, but because there were so few places to live and they could live there for free, some people made do. The top of Taree Bridge still stood proudly out of the water, but the road was usually underwater. Ron could recall what the town had looked like before the water covered most of it and thought he could find the old rowing club.

He sat there thinking. He wondered if, once he did find it, would they be able to get through the steel mesh and get the boats to the surface? He thought he could find the spot, but someone else would have to do the diving and heavy lifting. The trouble he had with his back and lung would make it risky, and if he died trying, it would be a burden on the rest of them.

They were all sitting quietly, looking at the fire or the sky.

Ron said, "The night is so perfect you would never know a disaster is coming. Have we decided what we're going to do?"

The women had been having a quiet conversation amongst themselves, and Sue said to the group, "We discussed it and think we should have a neighborhood meeting in the morning."

"Ok," said Ron, "there's always another morning, and we can deal with it then."

Ron woke up the next morning, fixed a cup of coffee, and turned on the TV news. There were pictures of floodwater pouring into the Grand Canyon from both sides. He couldn't see any landmass, just two waterfalls facing each other with the muddy water dropping into the abyss that was the canyon. The canyon walls were disintegrating, and the gap across widening. The foamy, muddy rapids were going over the sides and carrying the odd house or car with them. There were pictures of house roofs intact going over both sides. There was what looked like a small island with a tree, bushes, and the house with garage still standing, and the camera followed it until it disappeared over the side. Ron recalled seeing similar historic pictures, only they had been in China, and there had been no houses or cars. The coverage of the flood in China many years before had shown pictures of people standing on the sides of the canyon, watching the water pour over the sides, but there was no land to stand on in the pictures of the Grand Canyon. It looked like a sea separated by a waterfall, and anything that could float or wasn't tied down was being dropped into the abyss.

The news anchor said the amount of earth that was being carried away would eventually reach the Gulf and raise sea levels even more than previously predicted. He said the floodwater pouring over the sides was actually causing the sea levels to lower in the short term, but it was impossible to predict how much they would rise when the water reached the Gulf. There were also reports of more earthquakes and volcanoes in China, Russia, and the South Pacific, and pictures of spectacular eruptions in New Zealand. It looked to be chaos all over the world. Ron thought how it was becoming the norm. This time, it was China, Russia, and almost everywhere else, but it looked to be another beautiful day when he looked out the window.

After his coffee, he had hard-boiled eggs for breakfast. He used the salt and pepper sparingly, thinking they could become a luxury and hard to find in the future. Salt and pepper didn't have much nutritional value, but it could help him eat real food that might taste awful otherwise. He thought the same about honey and decided he should make sure to see his neighbor and stock up while he could. He thought honey was an ideal food because it wouldn't perish, and then he started thinking of the other foods that would be wise to stock up on if he could. He would need food he could carry with him and thought about how long different foods would last before they perished. He thought about all the chooks they had and how many they could take with them. They wouldn't perish if they could keep them alive, and they could use the eggs, but it would be difficult to travel with them, and he decided it was a bad idea.

He walked out the front door and looked across the street at neighbors already starting to gather around the picnic table. It looked as good as place as any to have their meeting, and he could see Hazel, Steve, Deb, Charlie, and Sue were already there. He walked over, and they were all smiles when they greeted.

Bill, the beekeeper, was there, and he was the first to address the crowd. "We heard there was going to be a meeting about the chooks and whether anyone or everyone wanted to go to Canberra or somewhere else, so let's get started."

Ron said, "We were talking last night because the news said we will be underwater soon and we could go to Canberra. Anyone who wants to come with us can, but they should have a boat, and we should all share the chooks before we leave. I think we should cook or salt as much meat as we can carry and split it up and sell the rest and divide the money. We can vote on it or something else if anyone else has another idea, but we need to decide before it's too late."

"Do you think we can get to Canberra by road?" Bill asked.

"We're hoping so, but we're taking boats, too, so if we have to, we can abandon the cars and get in the boats."

"Do you have a boat?"

"Yeah, I've got a twelve-foot tinny."

"What do we do when we get to Canberra?"

"Survive, for starters. I've heard Parliament House is safe now and is watertight, so even if it goes underwater, people inside can survive."

"That's alright for politicians, but will they let us stay there?"

"There are huge car parks underneath and conference rooms and the cafeteria as well as all the offices. It's supposed to be a safe place that can hold a million people, but right now, the suburbs there are well above water, so there shouldn't be that many people rushing to get inside."

"What about the longboats we were talking about last night?" asked Steve.

"If they are still under the rowing club and we can get them out," replied Ron, "they would be good to have. If there is a tsunami, maybe we could ride it out."

There was a lot of nervous laughter from the crowd responding to the idea of riding a tsunami wave on a long rowboat.

Ron said, "The longboats would at least carry more of us, and the oars might be better than a motor if there is a lot of debris in the water."

As he spoke, he was thinking out a plan and if it was better to look for the longboats when they were on the road heading to Canberra or before. If they could return home with the boats, they could load them up with supplies. If they couldn't find them or bring them to the surface, they would need to pack everything they had into cars, and those without boats could tow trailers. He brought up both ideas and asked people to think about them.

The discussion went back and forth, and they put it to a vote. They decided Ron, Steve, and Charlie would attempt to locate the boats and bring them back. Everyone else would start packing, slaughter the chickens, and gather as much food as they could and take as much as they could carry. They agreed to meet back at the same spot when it got dark and see where they stood, and the meeting finished.

Ron said to Bill, "I'll trade you what yabbies I have and as many cleaned chooks as you want for three of your kilo tubs of honey."

"I'll have to bottle all I have," replied Bill, and there's no point in leaving any behind. I'll give you more than that if you'll take it in a twenty-liter container with the wax."

"Heck yes. Leave it on my front porch, and I'll go grab the yabbies for you now, and you can take as many chooks as you think is fair from my share when they are divided up."

Ron walked back to the house, grabbed the bag of yabbies from his fridge, and went straight back across the road to Bill, who was still standing in the same place talking to Sue, Hazel, and Deb. Without interrupting their conversation, Ron handed the bag to Bill and then walked back home again. He went straight to his backyard shed and located a mesh bag containing his snorkel, mask, flippers, weight belt, and diving knife. He opened his toolbox, took out a hacksaw, and grabbed his bolt cutters from a shelf. He looked at his hand spear; he knew he wouldn't be fishing, but he thought it might come in handy. He could use it to check the depth up to ten feet, and it might be useful just in case there was a shark about.

He put all his gear into the back of his station wagon and went into the house. He had hard-boiled eggs in the fridge and grabbed them and a large bottle of water to have something to eat just in case it took longer than he hoped. He dropped the eggs and bottle of water into the back seat, and he was dragging his boat to the car when Charlie showed up as if on cue. There was no trailer, and it took both of them to lift the boat on top of the car and tie it down. As soon as they did, Ron got into the car and drove off, and Charlie went next door to Steve's to get into his car, his boat on a trailer, and followed behind.

The road to Taree hadn't changed much apart from a recently built temporary bridge and a few spots that were underwater. When the road was underwater, there were detours that went onto people's private paddocks on higher ground and then back to the road. It slowed the journey that used to be ten minutes to half an hour. Ron arrived at the riverbank and waited five minutes for Steve and Charlie to arrive. In the meantime, he had looked over the river and estimated where the rowing club used to be. He lined up the bridge and the roofs of the hotels that were sticking out of the water and had a reasonable idea of where they needed to go.

They put their boats into the water. Steve and Charlie also had a good idea of where the rowing club would be, and with their boat and motor, they were able to reach it first. Ron rowed his boat up to theirs, and Steve told him the water was clean enough to see the rowing club. He dropped anchor, and Ron tethered his boat to theirs. He told them he was hoping someone else would do the dive because he was a risk with a bung lung.

Charlie volunteered, and Steve told him to take the bolt cutters and cut the lock on the roller door first to see what was in there.

Ron said, "It's been years; I hope someone hasn't already beaten us to it."

"They wouldn't have locked the doors back up if they had," replied Steve,

Charlie put on the weight belt, mask, and flippers and went over the side with the bolt cutters in his hand. Ron and Steve watched as he reached bottom and cut the lock. Charlie came back to the surface for air and said it had been easy enough. He dove again and struggled to open the door under the weight of the water and mud that had settled at the bottom of the door. He came back to the surface for air and said the next part was a bit more difficult and then dove again. Ron and Steve could see he was struggling to pull something out from the door. He came back up to the surface for air.

"Here's something we didn't think about," he said."The boats are on a trailer, and it's full of water. How are we going to get it up here?"

Ron said, "The trailer has wheels, so we should be able to tow it along the bottom, and if we can get close enough to one of the cars, we can tie it on and use the car to drag it out."

"Take the anchor rope and tie it to the trailer," said Steve.

Charlie dove down again. When he came back to the surface, Ron untied his tin boat, and Steve started the motor on his motorboat and slowly motored away until the tension on the rope tightened as it met resistance.

Charlie said, "It's working. Keep going until it's all the way out, and let's see if we can get it to the boat ramp."

The rowboat rolled out easier than they expected and only balked when it went over the mud. Charlie climbed into Ron's boat, and they rowed straight to the riverbank. They tied the tinny up and walked down the riverside to the boat ramp because it was quicker than rowing. Steve towed the longboat along the bottom and headed to the boat ramp. Once he was moving, he was able to gain a little speed, and when Ron and Charlie were backing the station wagon down the boat ramp, Steve was there waiting for them.

They tied the longboat trailer to the tow bar on the car, and after the wheels initially spun, they were able to drag it up the ramp. When it reached the surface, they chalked the tires on the car and hurried to bail the water out of the boat. Then they were able to attach the trailer to the cars tow bar and drive it up to the road. They cheered for their success, and Ron suggested they should go back and see what else they could find, starting with the oars.

Charlie moved his boat close to the riverbank, and Ron and Steve waded out and climbed in. They motored back to the rowers club and took turns diving down and searching for anything useful. They used the bolt cutters to get through a wire fence and found a large steel box in which were life jackets and oars. There were two more longboats, but no more trailers. Without a trailer, they didn't think they would be able to tow the boats to the riverbank or get them out of the water any other way. The boats were too long to carry on the roof of a car, and without a trailer, they couldn't tow them behind the car either, so they settled for the one longboat and gear they had.

It was hard work getting their boats out of the water, lifting the tinny to put it on top of Ron's station wagon, and then hooking up the longboat and trailer to the tow bar. When they finished, they took a break and ate what food they had with them. Steve and Charlie followed close behind Ron just in case there was a problem with the longboat and some of the detours through paddocks. It was mid-afternoon when they started heading back, and they expected it would take longer to get home because of the longboat.

Ron, Steve, and Charlie rolled up to the home base just as it started to get dark. The rest of the neighborhood had begun the party, and they cheered when the trio parked their cars with the longboat on a trailer.

"Good job, but how are we all going to fit in that?" someone shouted from the crowd.

Charlie said loudly, "There's more just like that one if anyone wants to get them out, but there are no more trailers."

"There are also the smaller thin rowboats," added Ron, "and it looked like ten or so of those. I noticed the bar was still stocked and some of the whiskey, gin, and vodka bottles intact if you want to pick up something to drink while you're there."

Everyone laughed at the notion of diving underwater to locate bottles of booze, and then there was a low murmur in the crowd as they discussed whether or not they would try to get the boats out or take their chances on the road.

Ron said, "We're going to load this one up and take off at 8a.m.If anyone wants to follow us, we'll take you out there until you get the first boat out. You can use that to get the rest if you want, but we'll be moving on."

Bill said from out of the crowd, "Who is 'we' in your group?"

"Me, Charlie, Sue, Steve, and Deb," replied Ron, "and we can take Hazel. You can follow us if you have a boat. Otherwise, if you come with us, you will have to make do with one of the other boats. Six would probably be the limit in the longboat, or you might want to get one of the other smaller boats out. It's up to you."

There was a lot of grilled chicken and fresh produce. Everyone had taken every possible thing from their gardens, and they couldn't bring it all with them, so they were feasting. There was a lot of homebrew beer and wine that would be bulky to travel with, so a dozen or more of the men and some of the women were drinking as much as possible to keep it from going to waste. There was a large fire and the smell of burning hemp. It was a wonderful night that belied what was happening everywhere else around the world and the fear they all held for the next morning. The people in America were struggling with the huge floods that were pouring into the Grand Canyon from both directions, people in the Mississippi Delta were elsewhere or floating, there were earthquakes, volcanoes, and mudslides throughout Mexico and South America, plus volcanoes and earthquakes just about everywhere else in Europe and Asia. The tsunamis and nuclear disasters had just about wiped out Japan and Indonesia, the Philippines, Thailand, and Vietnam, and the world was in a horrible mess, but the Mid North Coast of New South Wales at that moment was enjoying a wonderful evening.

Just then, Deb came running from her house across the street. "They just put out the tsunami warning for the next seventy-two hours."

"Seventy-two hours?" said Steve.

"Yeah, but they are saying don't wait that long and get out now if you can. It should be twelve hours if and when it first hits, and then a second wave sometime after that, but they aren't sure. Then every twelve hours there will be another surge. The experts are arguing and admit they are guessing what will happen, but when they showed computer models, we will be underwater in all of them."

Ron said, "I'm going to pack up now and get some sleep. Then, first thing in the morning, have breakfast, and we take off. It should be around 7a.m. at the latest."

He walked over to his car, got in, and drove it with boat and trailer to his house, where he did a U-turn and parked on the street in front. He hurried into the house and started sorting what he could take, thinking of ways he could pack it tightly so it would take up as little space as possible. He packed a ten-kilo bag of dried lima beans and a ten-kilo bag of brown rice, all wrapped in plastic inside a plastic bucket with a screw-on lid, hopefully watertight. He filled the rest of the bucket with macadamia nuts stripped down to their inners shells. A few weeks before, he had been lucky to find three macadamia trees full of nuts that had dropped on the ground. He had filled a dozen ten-liter buckets and brought them home. He would pack as many as he could in any extra space he could fill.

He went into the backyard and started a fire under the barbeque plate and put a pot of water on top. While the water warmed up, he went back inside the house and loaded a duffle bag full of clothes, a blanket, and a pillow and put it all into the back of the station wagon. He went into the house and came back out with a plastic colander holding forty or fifty eggs and carefully put the eggs into the pot of the rapidly boiling water. He went into his shed and grabbed his old steel suitcase he used as a toolbox. There were screwdrivers, bolt cutters, carpenter's hammer, small sledgehammer, a rubber mallet, various sized shifters and open-ended spanners, pliers, crowbar, small tins of various-sized nails, a hacksaw, a handsaw, scissors, a block containing various knives, flint, and others odds and ends such as a magnifying glass and binoculars. *You never know what you might need*, he thought. He filled the spaces between the tools with more macadamia nuts because he thought them the perfect food, as they didn't need refrigeration or to be cooked.

He filled a bucket with water, raced back to the fire, grabbed the pot holders, and took the pot of boiling water with the eggs off the plate. He poured the contents into the empty colander, which he then filled up with hardboiled eggs. He dumped the eggs into the bucket of water to cool so they would peel easily later. He would fill his fifty-liter esky with as much food as he could cram in. He had large bags of avocados and oranges that would just about fill it. Then he thought the avocados would all ripen at the same time and the dried apples and dates would keep better.

He went inside the house and came out with his hand truck, loaded the steel suitcase of tools and the esky, rolled them out to the car, and put them in the boat. He hurried back through the front door and went straight into the kitchen. He took a screwdriver from the top of his cupboard and removed the handle from his fry pan; it would take up less room that way. He went into his bedroom and came out with a fifty-liter square plastic container and the lid. He put in the pan, two plates, bowls and mugs, two forks, butter knives andtablespoons, a fillet knife, two multi-tool pockets knives, and his favorite Smith and Wesson pocket knife. He threw in a kilo bag of coffee. He had bought two at a good price about a month before and hadn't finished the first one yet.

He put in a liter of salt and only then remembered the honey. He had traded the yabbies and chickens for the honey but hadn't seen how much honey there was. He went out to the front porch, where it was supposed to have been left while he was away. There it was, over in the corner, two ten-liter glass bottles of pure honey and a twenty-liter steel bucket with a lid. Ron took a screwdriver and opened up the steel bucket. It was three-quarters full with honeycomb wax and honey. He hadn't thought there would be so much. It could last him just about forever, but he would have to carry it. He wrapped the glass bottles of honey with large bath towels. He then took some bubble wrap he had saved over the years and placed it all in a wooden crate. He grabbed a bucket of macadamia nuts and poured what he could fit into the plastic container, and the rest he poured into the wooden crate over the top of the bottles. He put the steel bucket full of honey on the hand truck and then the plastic and wooden crates and rolled them out to the car and put them in the back of the station wagon.

The station wagon was just about full with the back seats down. The diving gear, fishing rods, tackle box, and a fishing bag were still in there from before. He moved the gear around and was able to make enough room to lift one of the back seats up. He would be able to take two passengers. He thought the food he had packed was enough to last three months, and the honey would probably last him ten years. He had enough clothes, blankets, pillows, towels, and soap. He took all the tools he thought might be helpful and wouldn't take up too much room. He packed a cast iron pot and fry pan he could use on a fire. It was certainly too much to carry in the car alone, but it would fit in the boat. He was hoping the others didn't have as much gear; otherwise, he might have to ditch something if they did need to abandon the car for the boat.

He was really hoping they would all be able to drive to Canberra, arrive with all their stuff, and set up a camp somewhere, and after that, everything would be great. At least they would survive. He might be able to get a paying job there or find a way to start a business. Maybe even find somewhere to play the piano, and all of a sudden, he thought this move to Canberra might be more than surviving; it could turn out great. He walked around the house for what he expected to be the last time. He thought about how the house would be underwater when the levels rose so he would need to swim around the house if he stayed. He laughed to himself at the notion of swimming room to room, and then he turned out the light and crawled into bed.

He lay on his back and thought of everything he had to do the next morning. Get a good sleep, wake up, shower, eat, and go get in the car. Drive it the twenty meters or so to Hazel's so she could throw her gear in. Steve and Deb, and Charlie and Sue would take Steve's four-wheel drive and tow Charlie's boat and follow them to the hospital, where he would drop Hazel off. She was too old and feeble to travel, especially if it needed to be in a boat. The hospital would remain above the water, and they would look after her as best they could. Then he would show the others who had followed them where the remaining longboats were, maybe help them get one out, and then they could manage the rest on their own. He should be on the highway to Canberra well before noon and arrive just after dark. It seemed a solid, foolproof plan. He closed his eyes and was asleep in seconds.

Ron woke up the next morning, rushed to the toilet, and after that, went straight to the shower. *For a clean start*, he thought. Then he made a mug of coffee and sat down to his breakfast of shredded wheat and cornmeal with four hard-boiled eggs. He finished his coffee and rinsed his cup and dishes. There was no reason to put his dishes away; he was leaving them behind, and they would be underwater soon, but he might as well take the cup with him. He would take the salt and pepper shakers too with the bag of ripe tomatoes that needed to be eaten. He had salt and pepper and other food packed, but he could take these things and use them and save having to unpack anything for later.

It was another beautiful morning. The sun was up, and there were no clouds, no wind and no birds flying around. Ron thought, *Are they sure a disaster is coming?* He got into his car, started it, and let it idle for a minute, and then he slowly motored up to the front of Hazel's house. He stopped, put it in park, and turned it off. With the weight of the boat on top and lugging the trailer behind and everything being loaded, it was a lot of weight for the little car. He would have to take it easy on the motor, or it would blow. Hazel opened her front door and waved. Ron got out of his car and walked up to her house. Just inside in the middle of the living room, she had placed two suitcases and a large carry bag. Ron took them one at a time and loaded them onto the boat. Hazel was standing near the car with her walker and handbag when Steve and Deb walked up and then helped get her into the front seat of the car.

Steve said, "We packed everything into Charlie's boat last night and are ready to go. I'll go with you and help get Hazel sorted. Deb will go with Charlie and Sue and meet us at the hospital. There will be a convoy of at least twelve cars, most with boats, following us to Taree, so we only need to show them where the boat club is and move on after that."

"Sounds like a plan," said Ron. "We should be sleeping in the ACT tonight."

He started the car, Steve and Deb had a few words and a cuddle, and Steve got into the back seat. Ron put the car in drive, and they were off. While driving to Taree, they commented about the beautiful, sunny day and tried to sound relaxed and unafraid. Hazel was a little bit worried about what would happen to her. She said they would only let her have two suitcases and a purse, so she'd put her purse in a bag and would say her bag was her purse.

"I've got my pillow and doona in there," she said, "and I don't know what kind of stuff they will have at the hospital."

"I don't blame you for that," said Ron. "I brought my pillow and blanket, too."

Hazel said they'd told her she wouldn't be there forever, just until things settled down and they could find somewhere for her and the rest of them to go. The hospital had advised it could manage seven hundred residents for emergency care and they would be determined by their age to make the elderly a priority. Ron thought there couldn't be seven hundred people older than Hazel around and she would be taken care of, so they could drop her off quickly and keep going.

He said, "Hazel, it should be good for you there, with a lot of people in the same boat, and you will make new friends, and you will probably all end up together in a little community somewhere."

She nodded but didn't speak .As Ron drove, he noticed that wherever there was water, it was lower than it had been. The first time he came across one of the little detours, he noticed it was a little lower, then a little more at the next one, and at the third detour, he could see the water moving.

"Steve," he said, look at that. It's like the tide is going out."

They pulled up to the hospital and helped Hazel out of the car and into admissions. Apparently, she was only the third person to sign in, and she was the oldest, so it was pretty certain they would have a bed for her. They said their goodbyes, and Ron said he would call the hospital to see how she was once they were settled in the ACT. Ron and Steve walked quickly back to the car. Charlie, Sue, and Deb were in Charlie's four-wheel drive, with his motorboat on the trailer behind and two kayaks on the roof.

Charlie opened his car door and stood up yelled over the top of the car, "We showed the others where the boats are, and we're ready to go."

Ron yelled back, "Let's go!"

He got into his car, started it, and slowly took off. When they reached an intersection and had to stop, he had a look at the river and noticed it was moving swiftly and looked dangerous. It was running lower and moving very fast, with circles spinning in the current. They were looking at the water and pointing at various things floating and spinning as they turned and headed the car around the roundabout and in the other direction. Steve was swearing. He looked at Charlie and Sue in the car behind and pointed at the river for them to have a look. Ron looked in his rear-view mirror and could see the backs of their heads, and as he drove ahead, he could see the other cars and everyone in them looking over their shoulders at the moving water as they went around the roundabout.

The convoy was moving along at a steady eighty kilometers an hour. They all noticed the water on the sides of the road was moving very fast. They were near enough to Newcastle to pick up a radio channel, so Ron turned on the radio. The news was breaking, and it was going to be an extremely low tide, then an extremely high tide, and they expected the change of tides over the following two days would be twice as extreme. They were advising people to get to higher ground now while they could.

Ron thought about what he should do. If he decided to keep going, maybe everyone would follow him, and they could be washed off the road. On the other hand, where would they find higher ground and would they be able to stay there for two days. He decided he would keep going, and if they came across a suitable place where they could all stop for a few days, he would stop and ask who wanted to stop and who wanted to keep going.

They were driving over the Karuah River. Ron looked out the window and noticed the water had slowed down considerably. He said, "We should make it to the ACT tonight and be safe there."

The traffic was starting to build, and they slowed down to sixty kilometers an hour. The convoy wouldn't be able to keep up if he tried to pass cars, and with his long trailer, it was hard to pass them anyway. There was more traffic for as far as he could see, and he figured he should just be happy they were moving.

An hour later, they were outside of Newcastle, and the breaking news was the tide was turning. They couldn't predict how high it would rise but advised the sea levels would be higher than ever before and would rise even higher over the next tide. The water pouring into the Grand Canyon and causing the flood in the Gulf with the rapid melt of what was left of the Antarctic ice was causing this massive change in tides and proving impossible for the experts to predict what would happen next. At the same time, they were reporting more earthquake, volcano, and tsunami warnings were just about everywhere. Ron looked around the sky to see as much as he could, and there was not a cloud anywhere. He remembered how some people used to say this global warming was nonsense and that some of the events were staged in some huge conspiracy theory. He thought that being outside on beautiful days like this had probably caused them to deny it, and then, when the business had hit the fan, it hadn't mattered what they had thought. He wished it had been a conspiracy theory; it would have been better if it was.

Three hours later, they were nearing Sydney. Ron noticed the water on the side of the road was again rising It was still a lot lower than the watermarks on the banks indicated the previous level had been, but it was noticeably rising. As they drove over a hill, they were met with a view of Sydney from a distance. The Opera House had been abandoned and was surrounded by water, but it was still there. The Harbour Bridge still had traffic and trains crossing, though there wasn't much land on the other side. Some neighborhoods and tall buildings built on hills had survived but were now on islands. Some neighborhoods were completely underwater, and some buildings that were standing were underwater to the second or third floors. It had been on the news that people were using boats to access those buildings and using them as squats.

The Sydney Opera House had been taken over by squatters before a concern of the Opera House had the police take it back. Then, because they could no longer open for business, they turned it into an organized squat and charged a small rent. People who lived there used a boat to get to the train. The piano was still there, and it was said that every night, the residents would gather around and sing and have concerts. They were a community of like-minded people. The Opera House concern was happy to get income from the rent, and some of the individuals who belonged to the concern were living there, too. Ron thought that would be a cool place to live, or maybe just visit.

The traffic heading out of town had been thick, but now there was a traffic jam up ahead, and it was standing still. There were flashing lights and police drones zooming around in both directions over the road. The drones carried a digital screen with lights flashing and text with audio speakers. The message: "Water levels rising, city is closed, evacuate to higher ground now." Then a little map would light up, with arrows indicating "higher ground."

Ron decided the next turn right was his best move towards higher ground. The traffic was moving a little faster, and as they approached the next turn, a drone hovered nearby with a flashing arrow pointing with the words "Higher ground." The traffic all turned the same way, and it seemed everyone caught on to the instructions as they continued to gain speed.

When the traffic slowed again, it was near a park with a huge parking lot, and there was another drone with an arrow pointing to higher ground. Ron pulled in, and at the side of the entrance were two cops.

One of the cops said, "You can park in here and be safe for the night. There will be a low tide in the morning, and you can continue out west then. If you keep going, you're going to be stranded on the road somewhere, or you will get washed off it."

Ron said to the others, "They say we will be safe here. I'll just park down the other end, then."

The cop waved, and Ron drove past. He found a vacant spot and parked, and the others pulled up right behind him. There were maybe four hundred vehicles in the area, most with trailers of some sort. There were soccer fields and other spaces filled with people in tents. There was music and noise in the distance, and the cars were still rolling in.

Steve said, "Lucky we got here when we did."

"Yeah," said Ron. "Let's hurry up and claim this patch of grass right next to us while we can."

He got out of the car, opened the back door to grab two blankets, and carried them over to the patch of grass. He spread out a blanket and then walked back to the car, climbed into the boat, and plopped the esky down to the ground. Then he climbed out of the boat and picked up and carried the esky to the blanket. The rest of them all gathered around and made their contributions of food and service to prepare snacks and get comfortable. People were walking past and saying hello or just nodding. The toilets were about fifty meters away, and there were several people in line outside the men's. There was a crowd gathering and some excitement on the road next to the toilets. A man was standing on the hood of his car and saying something to the crowd, but he was too far away for the group to hear what he was talking about.

The five of them, Ron, Charlie, Steve, Sue, and Deb, ate a large green salad. The ice in the esky was melting, and there probably wasn't going to be anymore during the journey, so they set out to eat as much of the perishable food as they could. There were cheeses, yogurt, rock melons, and grapes.

A young couple was walking away from the crowd, and the man was carrying a wooden box. They looked like hippies from the old days; the girl was very attractive, with thick, long black hair reaching her waist. The man had thin, oily, long light-brown hair parted down the middle and in a ponytail reaching the middle of his back. When they were near the group's blanket, they stopped.

The girl said, "Oh, look at that beautiful food. We'll trade you some wine for food."

"Sit down and join us," said Steve.

The box the hippie was carrying was a case of wine. The tall, thin fellow also had an untrimmed beard and mustache, and he was wearing wire-rim glasses. He held up a bottle of wine so everyone could see.

"We picked these up during the low tide. The liquor store had been underwater for four years, and today, we just walked in and dug boxes of grog out of the mud. It's still good stuff. We already tried a bottle, and we have cases of whiskey if anyone has something to trade and is interested."

The trade of wine for food seemed reasonable to both parties in their circumstance. They all sat around drinking two bottles of a wonderful red wine and eating the rest of the prepared food. During that time, the couple explained how they had known where to find the liquor store and how they'd known there would be merchandise intact. She had worked at the chemist next door for two years before the water levels had risen. The fact that it was inside a solid-brick building with steel roller doors was what had kept it all there. They'd had to use a crowbar to break in, and it had been a lot of work. There were other shops: a chemist, a jeweler, a cafe. They were all underwater again now that the tide was in, but they would be assessable at low tide if it dropped as low as it had that day. The couple said they were going to go back at low tide in the morning and try again. The tide was meant to be even lower than today's, so they could drive right up to the shops, load the car, and get out before the high tide came in. Even if they couldn't get back out of town, they figured they could stop for a night in a high-rise, and with so much jewelry and drugs to trade, they would be rich.

Ron said, "Not much good being rich if it kills you. Are people going to buy jewelry when they're just trying to survive?"

The young fellow said, "Yeah. There are no more mines or any new jewelry anymore. Rich women especially keep buying it, and when they do sell, it's at a profit."

Ron was thinking to himself that he wasn't going to risk going with them or even anywhere towards the city. Still, if something similar popped up with the low tide and wasn't out of the way, they should have a look. Several other people had stopped and were talking, and the possibilities of making a profit from salvaging items during the super-low tide made up much of the night's conversations.

The young couple traded a few bottles of wine for food and cash. The group sat around the fire with the new friends and talked for hours, smoking joints, joking, and laughing. Ron went to his car and fetched a blanket and pillow. He walked back to the edge of the blanket where his group and their new friends were, dropped his pillow, and then spread out his blanket next to it. He lay down on the other end of the blanket and then rolled towards his pillow, covering himself with the other end of the blanket like a hot dog rolled into a bun. He was on his back, with his head on the pillow, and he listened to the people talking as he fell asleep.

Next morning, Ron woke up in the dark. Everyone else was sleeping near the still-smoldering fire. He went for a walk to the toilet, and on the way back, the young couple who had provided the wine drove past. They waved, and the fellow leaned over and called out the window, "The tide waits for no one," and then they slowly drove away. Ron thought it could be fun to have a look around at low tide and see what they might find, and to visit the Opera House and play the piano or something else other than hurrying to get to Canberra before a hide tide wiped out the road.

He thought there were probably a lot of people thinking the same thing about fossicking through some shop when the water receded. So many, he thought, that maybe it wasn't such a good idea. There might be competition even if there were plenty of shops. Which type of shop would be the best to look at? Maybe he would find a hardware store and pick up new tools. Any electrical appliances would be ruined, but canned groceries could still be edible, and the jewelry and liquor stores probably were the most reasonable choices if he wanted to trade things for cash. The mud and water wouldn't hurt the jewelry, and it certainly hadn't ruined the wine they'd drunk last night. There wouldn't be much time to check it out and get back to high ground before the high tide came in, though.

The lack of time made the notion of visiting the Opera House rather dumb, too, unless he wanted to spend another night in town. The others wouldn't go for it anyway, and all going well, they could be in Canberra before noon. He could call his son when he got there and find out where the best place to camp would be. He could help the others set up camp and then probably spend the night at his son's and would get to be with the granddaughters. *That's a much better plan*, he thought.

He rolled up the blanket, and while the rest were still sleeping, he loaded everything he could into the car. He grabbed a dozen hard-boiled eggs from the dozens he had packed.

"Come on, everybody, wake up," he said."Breakfast is on me, hard-boiled eggs, and then we're out of here."

There was some moaning and stretching. Steve and Charlie stood up and walked to the toilet. The girls stood up, and each cracked an egg and peeled it. Ron handed them the salt and pepper, and they sprinkled some on and ate their eggs in two bites. When they finished, Ron offered them all another egg, and they all accepted.

When they were done, they each grabbed a small toiletry bag from out of their purses and walked to the toilets. Steve and Charlie walked back from the toilets, and Ron handed them hard-boiled eggs and pointed to the salt and pepper. He poured what water was left in the boiler onto the coals, and it made its last noise, gave a puff of smoke, and was out.

Ron said, "What do you reckon; we'll just go direct to Canberra but keep our eyes open during this low tide business just in case we see a jewelry store or something interesting on the way?"

"Sounds like a plan to me," replied Charlie.

The girls came back and got into Charlie's car, and Steve opened up the passenger door to Ron's car and got in. Ron and Charlie started the engines.

A lot of the other vehicles were also beginning to move. They pulled slowly onto the drive and followed a line of traffic onto the road. It was almost light, and as they drove along, they could see the tide mark from the night before was well above the road in most places. When they reached the bottom of a hill, the road was muddy, and they could see it would have been deep underwater and of course would be again at the next high tide. In Ron's mind, that verified his decision of getting straight to Canberra and not taking any extra time to look for treasure at low tide. It wouldn't be worth the risk. The traffic was moving along at 60kph, and in the daylight, they could see that the road heading in the opposite direction was closed. When they crossed a bridge, they could see the water was moving fast towards the ocean even though it was already freakishly lower than the previous high-tide mark.

Over the next hill, on the way down the other side, they could see at least a kilometer of bumper-to-bumper traffic. They slowed down to 20kph; it felt like a crawl, but at least they were moving. Then they stopped. Traffic didn't move for ten minutes, and then it moved ten meters or so and stopped again. It was stop and start until they were at the bottom of the hill, where it stopped again. There were no police drones or signs. After half an hour without moving, Ron and Steve got out of the car and walked up to Charlie's car.

Ron said, "have you heard anything on the radio?"

"There's a wreck ahead," replied Charlie, "and they don't know how long it will take to clear the road. Apparently, a ute pulling a boat on a trailer rolled over."

Just then, in the empty lanes that were closed in the other direction, the ambulance went screaming past, heading towards the wreck.

Ron said, "All we can do is wait it out."

They all got out of the other car and stretched. Ron climbed into the boat and took a jar of water out of the esky. He poured himself a glass, drank it, and asked who else wanted one. When they quickly drank the jar empty, Ron suggested they should have filled up a few more jars when they had the chance. There was another full jar, and they guessed it was enough, but they knew it could be a problem if they had to sit there all day. There was traffic as far as he could see in both directions. People were out of their cars and milling about. A few of them had set out lawn chairs next to their cars, and groups were gathering here and there.

An hour later, they heard the sound of car horns, and people were getting back into their cars. Two ambulances had passed on the other side of the road, and Ron could see two tow trucks in the distance, heading their way. It seemed the ordeal was over and they would start moving now. Charlie and the girls were already sitting in Charlie's car. Ron got in his car, and as soon as Steve got in, he started it up.

He heard some kind of noise he couldn't decipher, but he thought, *Who cares what it is?* He just wanted to get moving again. The traffic didn't move, so they were sitting there and idling. Something caught his eye to his side. He looked over, and there was a wall of water coming for them. There were cars and all kinds of debris floating and moving with the wave. He couldn't move, and there was nowhere to go if he could. It was such a shocking sight that he didn't know what to do and just looked in disbelief as the force of the water hit the side of the car. All the cars at the bottom of the hill were pushed off the road and carried away with the moving water.

Ron advised as loud as he could that it was time to hold on, and Steve nodded. It wasn't like they could have done anything else anyway. Water was coming in through the door, and it had already filled the car up to the seat. They watched the others in Charlie's car and saw Sue turn around and look back as they were all swept away in the new muddy river.

Ron's car had been moving sideways, but now it turned and was heading straight ahead with the current. They could see the ground ahead of them and the waters muddy rapids hitting it and taking whatever wasn't nailed down with it. They were moving now, not how they wanted or in the right direction but faster than they had been before they'd stopped. Ron thought about that and had a nervous laugh. The car smashed into a pick-up truck that wasn't moving until they hit it. Then that truck hit a tree, and it and their car stopped. They were stuck at a forty-five-degree angle on top of the truck's cabin. Ron and Steve were sitting there, looking up at the sky with their mouths wide open and not saying a word. The water and other debris kept flowing past. The boat and trailer were still attached, but the current was pushing it sideways.

They were looking at the catastrophe and the fear in each other's eyes when, at that moment, their boat was pushed off the trailer. A surge in the next wave, which was carrying a small car, had crashed into it. Then their boat went past them in the current.

Ron said, "Now, that's a worry."

He had been hopeful they were going to get out and into the boat, but now there was no plan. It and all the supplies in it were gone. They and their car were the only things above water other than the buildings and things on higher ground in the distance. They were still facing upwards at a forty-five-degree angle and had to stretch to look down and around. Steve had a wide-open mouth and look of horror but still wasn't saying a word. The water was roaring past them, and they were just happy to be above it even though they were stuck, held up by a truck in a tree.

Ron looked to the back of the car; it was underwater, and it was very unlikely he would be able to salvage anything. *Tools,* he thought. He opened up the glove box and took out his pocket knife and a multi-tool and put them in his pocket. He knew the toolbox would be at the bottom of the pile of stuff underwater and there was no point trying to get at it just then. It could wait, and maybe the water would subside, and it would be easy to just pick it up without getting wet. They weren't going anywhere for a while anyway, they hoped. In the meantime, the best thing to do was stay calm, slow down his heart rate and wait it out. That was a problem for Ron; his back was already aching, and sitting there for a long period of time would make it worse. Plus, his lung was hurting, and he was short of breath.

"Maybe this is the end, and maybe it's not," he said.

Steve nodded. "We're the only things above water as far as I can see."

It had clouded over, and with the flowing muddy water with a few brownish whitecaps and carrying muddy debris, it looked like the end of time, or their time at least.

Steve said, "I wonder what happened to the others."

"We don't know what's up ahead. We got stuck here. Maybe they were pushed up to higher ground and are all right."

Steve shrugged, and they sat quietly as the current moved around them. Every once in a while, a wave would hit the car or some debris in the wave would crash into them and make a racket, but after awhile, they relaxed as the world went past them. They sat still for more than an hour, and both were starting to nod off to sleep.

Ron said, "We might as well get some sleep if all we can do is sit here. One of us will have to stay awake, so if you want to snooze first, go ahead."

"I don't know if I can," Steve replied, but he closed his eyes to give it a try.

He had a sad look of despair on his face, and Ron thought to himself that he felt the same way. For a moment, he wanted to cry, but then he thought it was better to relax, remain positive, and save energy. They would just wait; they were ok for the time being and would be better off staying put and hoping for the best.

He sat there watching the water and the various things in it float past. The waves had stopped surging, and the current was slowing down. He sat there for another half-hour and noticed Steve had fallen asleep and was snoring. He laughed to himself and thought that was a good thing. He noticed the water had stopped moving and had become very calm. It was a lot less threatening now, but when he looked down, he could see the water level over the truck and that it would still be too deep to get out. That is, unless they wanted to swim, but there was nowhere to swim to. He could see land and houses up ahead and to either side. There were some people outside, but it was too far to see much or swim to.

After another hour passed, he noticed the water was moving back in the other direction. *It must be the tide changing*, he thought. Maybe it would recede enough and they could get out of the car and walk away. It was a lot to hope for, but it was about the only thing to hope for.

Steve woke, and Ron said, "It looks like the tide is going back out. Maybe if it gets lower, we can walk to higher ground."

Steve mumbled some obscenities and looked out his window. They sat quietly, looking out the window for another half-hour. The current was moving faster, and some of the debris they had seen flowing past them earlier was now floating past again in the other direction. Straight ahead, in the distance, they could see something large in the water heading in their direction.

Steve said, "What is that? Is that our boat? There's someone in it!"

There appeared to be five people in the boat. As it approached, they could see it was Charlie, Sue, Deb, and another rather large couple. Ron and Steve both opened their car doors and began waving. The people in the boat were waving back, and Steve was yelping. Ron looked into the back of the station wagon. The water had receded, and everything was covered with mud, but he could see the toolbox. He left his car door open and crawled over the back seat to retrieve it. He crawled back over the seat with the toolbox, checked his pockets to make sure he still had his pocket knife, and then stood up with his feet on the car seat and the rest of his body outside the car door. The boat was getting nearer, and everyone was smiling and excited to see them.

Ron said, "It's a minor miracle, but what else would we expect?"

The current had picked up speed. Charlie was rowing to keep the boat heading towards them, but it was obvious his control of it was tenuous. As they approached, the other man in the boat was at the front. He tried to grab the truck the car was resting on but couldn't keep hold. It looked like the boat would float past them in the current. At the last moment, Charlie took one of the oars and stretched it out to reach Ron. Ron grabbed it, and the boat stopped next to them. The current was moving swiftly, and they wouldn't be able to hang on forever. Steve climbed over the hood of the car and jumped into the boat. Ron started to choke up on the oar to bring the boat closer. Then he reached into the car with one hand and grabbed the toolbox from the seat. He threw it into the boat and then jumped to the boat. He managed to hang on to the side and climbed in. Everyone cheered. Steve and Deb were hugging, Ron and Sue had a little hug, and Ron and Charlie shook hands. Charlie introduced the new couple, Bob and Mary, to Ron and Steve.

Charlie said "We were swimming and thought we were out of luck. When the boat came floating towards us, we caught hold of it, got in, and then picked these two up a few minutes later. We have no chance of rowing against the current but can steer some going with it. Now all we can do is to try and find some dry land and get to it."

Chapter Three
A Day in the Harbor

They watched Sydney Harbour as it became larger and larger and could see people standing on the Harbour Bridge. Obviously, not everyone had evacuated the city. When a large wave suddenly appeared and hit the first row of buildings, they could see it breaking windows as it smashed into them around six floors above where the water line had been previously. They could see the water washing through the inside of the building where it had broken windows and could see debris washing around the sides at the same time. The wild weather at sea had picked up such force that a wave had picked up a cruise ship, and they watched as it crashed into a building. They guessed it must have been moored in the harborbefore the huge waves and high tide. The cruise ship stopped for a few minutes when it met the building. Then the entire building collapsed, and the ship crashed through what was left of it and carried on. The building or the part of the building above the water line crumbled and disappeared into the water.

People had been living in cruise ships. The cruise ships wouldn't voyage anywhere but had been considered to be a safe place to be when the water would rise. If there were people on this ship, they would've needed to be below deck because a lot of debris from the building had fallen hard onto the deck. The cruise ship was in front of their boat, and it was heading for the Harbour Bridge. Then a wave hit the bridge first, just making contact with the bottom of the bridge as debris immediately started to accumulate. Then the cruise ship hit the bridge and stopped there. The water line was above the bottom of the bridge, and the cruise ship's deck was above the road on the bridge. The current pushed the ship until it was sideways with the bridge, and then it appeared to be stuck fast. The arches of the bridge were just slightly taller than the ship.

Ron looked over at the Opera House. The doors were completely underwater, and only the sails were visible. There would be no way to get in, and he thought to himself it was a good thing they hadn't tried to go in for a look before, or they would probably be dead now. Even though they were in trouble, at least they still had hope, and it would be worse for anyone in there now. Then a large section of the recently felled skyscraper was floating atop the water and crashed into the Opera House. They watched as a sail was ripped off the roof of the Opera House and pushed into the current. The sail was floating swiftly with all other debris. It seemed to hold air for half a minute, and then it collapsed and was covered up by the rest of the debris. There were still people inside the Opera House, and they were now sitting in open seats without a roof. They could be seen crowded in the upper seats, with the water just meters below.

Then another container ship crashed into the cruise ship that was stuck against the bridge. Both ships were held fast, with the cruise ship between the container ship and the bridge. There was the sound of steel screeching as the ships slid and rubbed together before they all came to a complete stop under the bridge. There was another massive noise as another building crumbled into the sea. There were people swimming and hanging on to debris as the current sucked them towards the bridge. When the people swimming reached the ships stuck on the bridge, some managed to climb onto the ships, and some climbed onto the bridge.

Their long rowboat was speeding ahead and stuck in the current, and none of them had seen another container ship approaching from their right. The ship was listing, and their boat slipped over its side and onto the deck. They barely had time to scream before their boat slid along the deck of the ship with several containers that had broken free from their hold and were also sliding around the deck. It was another minor miracle that required no effort on their part and saved their lives.

The ship they were hitching a ride on eventually collided with and held fast to the other container ship on the bridge. Everyone in the long rowboat was safe, but their boat was stuck on the deck of a ship that was stuck with two other ships against the Harbour Bridge. The tide swept past them, carrying all types of debris, cars, trucks, parts of buildings, and anything that could float as they sat silently with wide-open mouths and eyes.

Ron said, "You won't see this in the tourist brochures," and there was some quiet laughter. "We're just spending a day boating in Sydney Harbour; where else would we want to be?"

"I can think of a few better places," said Charlie, "but there's a few worse places right now."

"Yeah, the Opera House doesn't look like it would be much fun now."

They had a look around. The other people in their predicament on the other boats were doing the same. They waved, and Ron and Charlie waved back. There were people on the Harbour Bridge looking at them, and they waved at them, too. Then they could see people on the cruise ship looking out of their windows. They were all on the top floor and waved when they noticed everyone else looking around and waving at folks. There were survivors all around. but none of them could go anywhere, and they were all still vulnerable to the disaster.

Just then, it started to pour rain. Ron thought maybe they should have tried to climb across and get on that cruise ship; they would at least have been out of the rain. It might have been dangerous to get there, though, and the people inside might not want to share the dry space.

"I've got a tarp under here," he said. "Let's get it out and try to stay dry."

He picked out the tarp, unfolded it, and they tied it to the rings on either end of the boat, with enough room for them to get under it. The sound of the heavy rain was getting louder, and they raised their voices and used gestures to communicate as they sorted out their places to get as comfortable as the conditions would allow.

Ron said, "We may as well get some sleep while we can. We have three pillows and blankets, and we'll have to take it in turns."

The couples all cuddled up and shared a pillow and a blanket. Ron put on a jacket and picked out a jumper to use for a pillow. He volunteered to take first watch so the others could sleep. He was awake for a few hours, but he closed his eyes, and a few minutes later, they were all fast asleep.

The rain was pouring, and all seven people in the boat were snoring. They were all drying and snug while the chaos outside the boat carried on. There was noise from the rain and the sound of the current running into the ship and welling up as it washed around, rocking the boat slightly. It was a perfect time to sleep: with the sounds of the water and while they could do nothing else.

After a few hours, the rain stopped. The water that had been rushing so hard had settled, and everything went quiet. Ron woke up. He saw the rest of them sound asleep and remembered where they were. He listened to the quiet and wondered how long he had been asleep, and then he rolled over and peeked out from under the tarp. It was still cloudy but much lighter. He looked over at the other two boats that were on the deck. All the people were gone. He looked up to a window that must have been the captain's deck and saw some people looking back at him. He recognized them from the boat as they waved. He figured they must have scurried out of their boats and gone up there to get out of the rain. He thought maybe that was what they should have done, too, but all was well enough. He waved back and looked up to the cruise ship and bridge and saw people were still there, too.

Those on the bridge must have been soaked, and there wasn't as many of them as before. They were moving about, and some were walking down. One guy in his underwear was wringing out his clothes. The water in the harbor was brown and still with debris floating on top. The people who had been in the now open-air bleachers at the Opera House had moved under cover, or so it appeared. Without the sails, the Opera House looked like several rows of bleachers on top of the water. Then he noticed a body floating with the debris, and another, and another. He wondered what he should do: try and retrieve them or something else. He thought he wouldn't be expected to do that; it would be a huge risk, and it was best to leave that for the police and emergency services. The emergency services were going to be very busy, even overwhelmed, and that was just the ones who had survived or didn't need rescuing themselves.

He looked over the ship to see if there was a way they could climb out. Maybe they could climb over the ships and onto the bridge and just abandon the boat or maybe push it off the ship and row towards land and Canberra. The climb looked difficult, and it would be impossible to take the food and other stuff. Staying right where they were on their rowboat on top of the ship was the best option. Maybe they would get rescued and be flown to Canberra in a helicopter. That would be the best for them, but it was just wishful thinking. Then he heard a helicopter and looked up to see it approach the Opera House from the other side. He thought it was probably a news crew doing a flyover, and even if it was a rescue, there would be others in dire need.

Then, without warning, another building collapsed and folded into the harbor, breaking the silence as metal and glass hit the water. Everyone woke up, and Ron heard voices. He looked back under the tarp and told them everything had gone calm before and that the noise was just another building collapsing. Charlie was untying his end of the tarp, so Ron untied the other, and they lifted the tarp off to the side. Everyone was on their feet, and they helped fold the tarp until it was a small square, and Ron placed it under the hockey straps and over the food. They also folded up the blankets, and Ron put them and the pillows under the tarp.

Ron said, "What should we do now?" There was no response. "I'm thinking we should drop our boat in the water and row towards land and find our way to Canberra by road."

"We can't stay here and wait to be rescued," said Steve. "It might not ever happen."

"We should probably go now before the tide goes out again."

"When is that?"

"I don't know. It could be soon, or three to four hours at the most. If it starts going out again, we should latch on to something or land so we don't get caught up and dragged out to sea."

"Ok," said Charlie, "let's go," and everyone else agreed.

Ron, Charlie, Steve, and Bob got out of the boat and dragged it to the edge of the ship's deck while the girls stayed seated inside. Ron called for them to put on life jackets and tie them tight. The ship was leaning, but there was still a fifteen-foot drop from the deck to the water. There were also three dead bodies floating among the debris in the vicinity they intended to drop the boat. One of the girls seemed to be calling for God, and there were whimpering gasps from the others.

Ron said, "You girls will have to get out; we'll drop the boat in the water. Then jump into the water next to the boat and climb in. We better jump in one at a time."

The girls climbed out onto the ship's deck, and they positioned the rowboat sideways so that it was sitting lengthwise and leaning against the deck's handrail. Ron made sure the oars were secure under the hockey straps and declared the boat ready. They lifted the side of the boat over the side on the ship's deck, lowered it as much as they could, and then dropped it overboard.

Ron jumped into the water straight away, grabbed onto the side of the boat, and climbed in. Then he looked up and called Charlie, and Charlie jumped in the water and landed next to the boat. Ron helped him climb in, and the boat was moving away from the ship. Ron took the oars and handed Charlie one, and they rowed the boat close to the ship. Steve jumped into the water, and Ron and Charlie both helped him climb into the boat. The boat had drifted again, and they rowed up next to the ship again. Ron kept a hold of the oars and tried to keep the boat near the ship as Bob jumped in and Steve and Charlie helped him climb in. Then they went through the same process for each of the girls, and it had taken them half an hour to get everyone in the boat and ready to row.

They rowed around the end of the ships and then went under the bridge and towards land. The current had been stagnant but was now starting to provide some resistance. Ron and Steve looked at each other with concern and rowed harder. They looked back at the Harbour Bridge; it was about a hundred meters behind them, and ahead, all they could see were buildings standing out of the water and seemingly endless water past that. There was a single building about fifty meters to the right.

Ron said, "Let's aim for it. We can latch on and then see if we have any better options."

Steve nodded, and they moved toward the building. The current was definitely getting stronger, and debris was moving with it. Every so often, a wave would rise and crash over the receding tide. They rowed to within ten meters of the building. The current was swirling circles on top of the water and started to push the boat sideways.

The others in the boat had been sitting quietly since they had gone under the bridge, but the sideways nudge had them up and talking. Bob and Charlie crawled over to help with the oars, and the girls went and sat at the front of the boat. The girls were trying to reach the building, and they touched it, but there was nothing to hang on to. The waterline was probably at the third or fourth floor, and the glass windows offered nothing for them to latch onto. The current was moving faster and faster, and the four men were struggling to keep the boat close to the building.

Ron said, "We'll have to break a window or something. It's amazing they all aren't already broken."

The wind and the current pushed their boat sideways and up against the building. Ron took an oar and swung it into a window. The oar bounced off it. Bob took the other oar and did the same. They each had another swing, but the windows wouldn't break. Then the raging current had taken them past the building, and they were rapidly drifting away. They would have to row against the current to reach the building. Ron and Bob quickly placed the oars back into the oarlocks, and they were immediately rowing again, but they were thirty meters away before they started to move.

They managed to keep the boat steady in the one spot for several strokes. Then they heard a loud noise. None of them knew what it was, and they looked at each other and around. It was the sound of twisting or bending metal and then some pops and bangs, and they turned to look in the direction of the sound. They saw the windows being broken as the building they had just tried to latch onto began to collapse.

Ron said, "The windows are broken now, but it's a good thing we're not in there."

While it collapsed, it looked like slow motion at first, and then the debris was falling on them. There was dirt and grit all over them and on the boat, and they were hoping nothing heavy would land on them.

It was because they had stopped rowingwhen they looked around that probably saved them, as the current had taken them just far enough away from danger. There was a surge in the water when the building was collapsing that pushed them along further, and then a large bit of debris fell on the front tip of the boat, just missing the women. The force was such that it tipped the boat almost straight up and down on its nose, and all the men were thrown into the water. The women hung on as the boat tipped to its front end and then dipped under, and as the back end went up in the air, paused for a second, and then crashed safely back into the water.

The boat kept moving in the current as Steve and Bob managed to swim to it and hang on to the side. Charlie had fallen on something hard that had been floating on the water. He put his hand up, and it was obvious he was struggling. He hung on to what looked like a timber wall frame and was moving with the current. Ron was trying to swim toward the boat, but it was taking him too. Deb and Sue helped Steve and then Bob climb into the boat as Mary tried to hold it steady with the oars. That took a few minutes, and both Charlie and Ron were being swept further away.

As soon as Steve and Bob were in the boat, Sue went to help Mary row. Ron watched as they moved toward Charlie. He also saw other people swimming. He guessed that they had fallen out of the building and survived, but whatever the reason, they were there in the water, too. He could see Steve and Bob lift Charlie into the boat. Charlie looked like deadweight as they dragged him over the side and into the middle of the boat.

Ron started looking for something to grab onto. With all the debris floating around, he thought it would be easy to find something to help him stay afloat, but he couldn't find anything useful. He was tired, and he stopped swimming and let himself float in the current. He was watching the boat become smaller as it drifted further and further away. He put his hand up so they could see him. He saw them stop and help one of the other people who were in the water into the boat, and then there were others swimming towards the boat. He made another effort and tried to swim towards them, but the current was taking him in the other direction faster than he could swim.

He was floating and relaxing when he noticed the current had slowed, and only a few minutes after that, he noticed the tide was running out and he was going with it. He knew he couldn't swim against it and would become too tired and drown if he tried. He turned on his back and floated while he tried to catch his breath. His life jacket was making it easier to keep his head above water, and without it,he would have drowned, but it hadn't helped him swim against the current. He relaxed and the let the current take him.

He tried to keep an eye out for the boat from time to time, but eventually, he gave up. He thought he should be floating under something pretty soon and would be able to hang onto something there and climb out of the water. He'd lost sight of the boat, so he kept his eye on the bridge. He noticed the cruise and container ships were still there, but they were now beginning to separate from the bridge. They were drifting apart and slowly being taken back out to sea. He remained calm and started paddling towards the bridge, and he found it getting easier to move in the direction he intended.

The tide appeared to be going in both directions, with a hundred-meter-wide series of millions of whirlpools separating the rapidly moving waters. At the same time, large waves were forming in both directions, and compared to the waters moving in opposite directions, the water with the whirlpools appeared calm. He found a piece of Styrofoam floating past and grabbed it to use as a paddle board, and he made his way to the calm water and swam towards the bridge. He thought it must be a tidal wave crashing at a very low tide, if that was even possible. By the time he was near the bridge, the ships were at least fifty meters away and heading back out to sea. Most of the people who had been on the bridge were gone, and the water was running about five meters under the bridge. He reached the bridge and then let go of his piece of Styrofoam and grabbed onto a pillar and held steady.

He was holding onto the corner of one of the bridge's pillarswith his feet out of the water and a hand on either side, trying to dig his fingers into the stone. He couldn't find anything better to grab onto and was finding it a huge effort to stay put. He looked for a way to climb up and maybe reach the road, but it looked hopeless. He hung on and took a few deep breaths.

Just then he heard Steve yelling at him to "Hang on!" He looked up and saw the boat heading towards him. There were more people in the boat, a rotund couple and an Asian couple with two young children, a boy and girl maybe eight and ten years old. There was a swimmer in the current heading towards the bridge who put his hand up in a gesture for them to help him. They motioned and gestured back that there was no more room in the boat, and the swimmer turned and continued swimming. Steve was rowing the boat, going with the current but aiming for Ron on the bridge. As they approached, it became obvious the current wouldn't let them stop and they would be drifting past the bridge. When they were as close as they were going to get, about five meters from where Ron was hanging on, he jumped in and swam to the boat, just managing to hang on to the side.

One of the new passengers said, "There's not enough room."

"Help him in," said Charlie."This is his boat."

The new guy had a puzzled look, but he reached out and grabbed Ron's arm, and Steve crawled over to help him hoist Ron into the boat.

Charlie said to the newcomer, "Well, the boat is all of ours, but most of this stuff in it, like the food, he put in here."

The new guy shook his head and then held out his hand, and Ron shook it.

Steve said, "Mark, this is Ron. Ron, this is Mark and his wife, Amy." Ron shook Amy's hand. "This is Jin and his wife, Julie, and their children, Jessie and Eddie." Steve motioned to the Asian couple and their children.

They waved, and Ron waved back, and there was some hooting and cheering from the others.

Deb said, "We thought we might not see you again."

"Me, too," Ron replied."So, there will be thirteen of us in the boat now."

"It will be crowded, said Steve, "but we'll manage."

As they were talking, the current was taking the boat further out to sea, and they were helpless to stop it. The bridge was behind them and getting smaller.

Ron said, "We should try and latch onto something big, but if we can't, we can hang on until the tide washes us back in."

"Let's hope it does," said Charlie.

Steve and Bob were rowing and making slow progress through the current. They were aiming to get back to land, or at least to keep moving in that direction. They had caught up to the cruise ship that had been stuck on the Harbour Bridge and were looking for a way to latch on. They couldn't expect they could be so lucky again to have a ship list enough in the water so that they could just land their boat on deck. Now that the ships were upright, Ron realized what freaky luck it had been before. He thought that particular type of luck unlikely to happen again, but since they had been so lucky so far, maybe it could continue, and they could find something to latch onto. He didn't want to be taken further out to sea and was sure none of the others did either.

Ron and Mark made their way to the back of the boat, climbing over everyone else, and helped Steve and Bob with the oars. Charlie thought he might have broken ribs when he'd been thrown out of the boat before, and it was obvious he couldn't help. The rest were using their arms over the side of the boat, trying to help them row and doing everything they could to keep moving in the right direction, but they were in another rip and found themselves unable to row against it. There were expressions of despair when they stopped rowing.

"All we can do now is sit and hope for the best," Ron said.

It was cloudy and almost dark as they floated out of the harbor, and as they were taken out to sea, the Sydney sights became smaller and smaller until it was completely dark.

Ron sighed. "If anyone can sleep we may as well, there's nothing else we can do."

There were thirteen people in a long rowboat but only four seats, and they were being taken further out to sea. The two children and the mother were sitting on top of the tarp. Sue, Deb, and Mary were sitting on the seat in front. Jin and Amy were sitting on the next bench, Steve and Bob were next, with the oars in the boat, Charlie was on his back on the floor between the next bench, and Ron and Mark were at the back of the boat. A set of waves washed over them and continued to drag them further out to sea. They would have to do everything they could to keep the boat upright and stay in it before anyone would be able to sleep.

The dark was over them, and they had no idea what time it was or how far they were from shore. They had been so busy rocking up and down in the boat and just hanging on trying to stay in it and upright that they didn't think about time or distance. It must have been four or more hours later and was still completely dark when the boat's rocking slowed. The children and Charlie were sleeping. Ron picked up a flashlight he had packed and stowed on the floor under hockey straps. He turned it on, suggested some of them should get some sleep, and made a motion with his hands folded under his head to Julie. She made the gesture back and then held out her hand to Jin, and they lay down with the children. Ron held the light on them and then the girls. The girls sorted themselves between the bench and floor. Ron shone the light around the boat, but there was only water.

He asked Bob and Steve if they wanted to lie down next to Charlie and get some sleep, and they both said they were ok and would stay awake. Ron asked Mark if he wanted to sleep on the floor or the bench, and Mark slapped his hand on the bench, so Ron lay on the floor next to Charlie. He fished his jacket out from the gear under the hockey straps, turned off the flashlight, and put it back under the hockey straps. He folded the jacket to use as a pillow and said good night.

The next morning, the clouds were blocking the sun, and the fog was thick in the air. The sea had calmed, and everyone in the boat was sound asleep. Ron woke up first and looked over the boat. Bob and Steve were on one bench with their heads on either side of the boat. They looked rather uncomfortable and like they would be sore and sorry when they woke. Mark was lying across the bench that wasn't wide enough to straighten his legs. Charlie was on the floor and couldn't stretch his legs either. The girls and the family were all close together and partially on top of one another.

Ron thought they must all be worn out to be sleeping in those conditions. They were going to be much more miserable if they had to spend another day and night in the boat. There wasn't enough room for thirteen people; it would have been tight with the seven of them. At least the seven neighbors they'd started out with had survived, and that was a minor miracle as it was. His back was sore, and his leg muscles were tight. He stood up and stretched, reaching for the sky, and then he pushed and arched his back. His body felt like it was completely bruised, and every movement was slow. He moved his legs and walked in place for several paces. He stretched his arms up to the sky and from side to side and searched the horizon on all sides, but there was nothing but fog.

Charlie opened his eyes and said, "Where are we?"

"In the fog," Ron answered. "Can't see a thing."

Steve and Bob popped their heads up and had a look. The rest of them didn't stir. It had been a harrowing experience, and the sleep was doing them good.

Ron said, "Charlie, are you alright?"

"I think my ribs are broken, and I smacked my knee," Charlie replied.

Steve and Bob were stretching and rubbing their eyes. "Should we start rowing?" asked Steve.

"Which way?" said Ron. "We might end up in New Zealand."

"New Zealand might be closer," said Bob.

They laughed lightly, and Charlie said, "New Zealand probably isn't even there anymore. They were having earthquakes and eruptions. The waves might have wiped them out."

"It sure has changed things again," said Ron."We may as well drift instead of taking a chance of rowing in the wrong direction. I have some dried fruit we can have this morning. Then we can soak some beans and have them tonight in case we don't hit land."

"Don't say that," said Steve."We don't want to be out here another night."

"No, we don't, but we may as well be prepared if we can."

Everyone went quiet for a few minutes. Ron climbed over the next seat and fetched the container with the food from under the hockey straps. He asked if anyone wanted dried figs and dates, and they all put up their hands and said yes. The children and then their parents woke up. Ron picked out a small handful of dates and handed them to Charlie, who stayed on his back. He handed some to Steve and Bob and then climbed over the next seat with the container and opened it so the children could grab some. Then he offered it to the parents, who took some and then bowed and said thank you. The children then said thank you. Deb woke up and peered over.

"Do you want something to eat?" Ron asked."We've got dried figs and dates this morning."

"Yeah," said Deb."Oh, yum."

The other girls woke up and sat upright. Ron offered the container, and they each took a handful of dates. He sat there and ate a few figs and dates, too. Mark woke up and asked what was happening, and Ron told him nothing much and offered him something to eat. He offered the container to Mark, who took a handful of dates and then a few figs.

Ron said, "I was going to soak some beans for later, but we may as well eat the rest of the hard-boiled eggs if we need them if we're still here tonight."

He was hoping they wouldn't be out there for another night. He wanted to be on land, and he also wanted to save his food. It was supposed to last him for weeks or months once he got to Canberra, or so he had thought. He took a handful of dates and motioned to Mark to pass them around again. Mark took another fig and passed the container on. Everyone had another few figs or dates, and the container came back. Ron thought about what would happen if they were lost at sea for more than another night. He looked inside and thought there would be enough dried fruit for them to have one meal of it a day for three or four more days. The beans and rice could provide them all another meal per day for several days. He thought that if they could have made it to Canberra that day, it would have fed him for months.

He thought some of the others might have food. Steve and Deb and Charlie and Sue had all brought backpacks that were still under the hockey straps. The newcomers looked to have nothing. He thought they would have to share food equally, as it would be too awkward to eat in front of the others especially if they had nothing and were hungry. He decided the best thing to do would be to find land before the end of the day, and then everything would be great. A bottle of water was passed around, and everyone had a cup. Ron looked at the amount of water they had and knew that it would be the bigger problem. If they were going to soak beans and have water to drink, the twenty-liter container and three two-liter bottles they had with them would be gone before the food.

They drifted in silence or quiet conversation for hours as the fog lifted, and the sun began to shine. The surf was gentle, and the morning sun was heating them up. There was no wind, and the sea was on the horizon in every direction. There was also smoke on the horizon in every direction. Conversations started all around the boat about which way Australia was and what would be happening there after this latest disaster. Jin and Julie had been the last ones to see the news. They told how all around the world, more earthquakes and volcanoes had hit or erupted the previous day just before the tsunami had hit. Italy was almost completely destroyed, and the Rock of Gibraltar was an Island. The French Alps was supporting life, as were other mountainous areas in Europe, but nuclearfallout was affecting most of the population there and was a threat to the rest.

The floodwater pouring into the Grand Canyon had subsided, although now there were several small rivers from the East Coast meeting the Mississippi, and from there the Grand Canyon. The Colorado River at the bottom of the canyon had receded to expose cars and other debris in huge piles. Most of the American Midwest had been flooded and was now a wasteland. Whole cities, homes, and buildings had been destroyed. The floodwater had only left debris covered in mud. People were surviving on higher ground but also living without electricity and services. South America had been largely devastated, but the population had moved to higher ground. It appeared the Americas had become thousands of Islands, and that had been before the latest surge of volcanoes and earthquakes. Japan was reported to be uninhabitable because of the nuclear fallout and higher seas, as was most of East Asia.

The tsunami warning had been made the day before it hit and was expected to remain in place for several days, meaning they expected more than one tsunami. Ron had another look around the horizon: barely a breeze, calm seas, and blue skies for as far as he could see. He thought that was becoming common. All over the world, there were wild weather events, volcanoes, and earthquakes, but where they were, it was calm and perfect. Apart from the tsunami, of course, but it didn't look like it would be an issue anytime soon. He wondered how long they might have before the next tsunami and if they should they start rowing. It could be rowing would get them to shore sooner and maybe just in time, or maybe rowing would be wasting energy for nothing.

He said, "I think we should take turns rowing, I'll go first." He pointed to the sun. "We'll just row away from it and chase our shadow." He took the seat with the oars and started rowing.

Steve sat down next to him. "Here we are. Give me an oar."

They each had an oar and in quick time were rowing in rhythm. The sea was calm, and it seemed they were making reasonable progress. Ron said he hoped they were going in the right direction, and Steve nodded. That was all they said for a good twenty minutes as they kept the pace steady. Suddenly, Ron stopped and said he needed a rest, and Steve immediately agreed. They caught their breath for a few minutes and then rowed at a slower, steady pace for another twenty minutes. They looked at each other and stopped rowing without saying a word.

"Anyone else want to row for awhile?" asked Ron.

"Look over there," Mark said, pointing.

Off to the left and ahead, there were hundreds of birds working the ocean.

"Must be something," Mark said, "and it's almost on the way. Maybe we can jag a fish, and we can have sushi."

"It could be anything, if you want to have a look, maybe you could row over there," said Ron.

"Yeah, we should." Mark motioned to Bob.

Ron and Steve changed seats with Bob and Mark. Ron stretched his arms and arched his back, and Steve did the same before they sat down. Bob and Mark started rowing towards the birds.

They rowed at a slow, steady pace for around half an hour. They were still a distance from the birds but could see what they were doing. There was a huge pile of debris that obviously had attracted fish, and as a result, the birds were there, too. When they were near enough, they could see the objects in the debris. There were timber frames from previously cherished homes, a lot of sheet metal, plastic chairs, and other things bobbing up and down. There was what appeared to be small islands of green bushes and one with a small tree. There were four legs sticking straight up out of the water that looked to be a dead horse or cow. The birds were all over it. Some were diving in the vicinity and were active, to say the least. Ron thought, *That's what birds do. They take advantage of a situation while they can. There's no emotion with them. To them, this catastrophe is a bonanza.*

As they neared the mass of floating debris, they noticed sharks were also around in numbers. The sharks were concentrated in several spots and also appeared to be very active. They were moving fast, and every so often, they would thrash about on the surface.

Ron said, "There are probably more dead animals in there attracting the sharks. Maybe we better not get too close after all."

Bob and Mark stopped rowing, and everyone was silent as they watched the birds and sharks working for a feed. They could see all kinds of things in the floating mass of rubbish. Most of it was the typical trash, plastic bottles and such. There were other items that appeared to be walls and roofs from houses and pieces of sheet metal.

Ron spotted a few plastic buckets and thought they might be worth picking up. There might be something useful in them, or at least they could use the empty bucket for a toilet or something.

It was kind of dicey going to the toilet over the side of the boat. He noticed some of the sheet metal was nailed to timber as well, and there was other timber they could have used to make a fire. He saw a child's plastic three-wheeler pedal toy and thought that could go to one of the kids, except that was a dumb idea while they were all in a boat.

"What do you guys think?" he asked. "Do you want to have a look or stay clear of it?"

"Stay clear," said Charlie.

"Yeah," said Steve.

Bob and Mark agreed, and Sue screamed, "Oh, God," as a shark came to the surface near her at the front of the boat. "Let's get out of here."

Bob and Mark started rowing away from the rubbish as sharks darted around the boat.

Ron said, "I don't think they will attack us when they have all that other food to eat."

"We'll get out of here anyway," said Bob.

They started moving much faster and kept the speed up for around five minutes, and then they stopped rowing. Everyone looked around the boat and back at the floating pile of rubbish. The sharks appeared to be gone, and Mark asked if anyone else wanted to row.

Jin and Julie volunteered and changed seats with Mark and Bob. The children changed seats and sat with their parents. Ron suggested they all have a drink of water, and he took out the container and poured a glass and passed it on. When the glass came back, he poured another one until everyone had drunk a glass. Then he poured himself a glass and put the lid back on the container.

Jin pointed and said, "That way?"

"Yeah," replied Ron. "Let's go."

Jin and Julie started rowing at speed. They kept it up for several minutes, and then Julie stopped. She asked Jin to slow down, and he smiled and nodded, and they got into a slow, steady rhythm. They rowed for around an hour and then stopped. Nobody said a word. The children were lying down on top of the gear on the floor. The girls at the front of the boat were sleeping, as was Charlie and Steve. The rest were awake but sitting very still. A few minutes later, Jin and Julie started rowing again. They kept a slow, steady pace for another hour and then stopped again.

Ron said loudly to the back of the boat, "Do any of you girls want to row?"

Deb gave a negative grunt, while Amy looked at Sue, and they made a gesture to indicate they weren't keen on rowing. Steve stood up, and Ron and Steve swapped places with Jin and Julie and started rowing. They rowed for around half an hour. Then they stopped for around five minutes and went again for another half an hour and stopped.

Ron said, "We should think about cooking up some beans before it gets dark. It will be easier if we have light. Twelve cups is about all the pot will hold, and it should be enough for all of us."

He was keen to use his little grill. He didn't have any gas or pellets, so they would have to burn wood. He had two small bundles of sawed wood he had picked up from the old chook shed and packed in the boat. One bundle would be adequate to keep the fire going long enough for the beans to cook. The barbeque was rather small, so it would have to be a small fire, and he would have to keep feeding it. He had used it at home in the backyard with no problem, but it might be a different story on the boat.

"What do you guys think? Should we stop and start a fire to cook some beans?"he asked.

"Yeah," said Steve, "if you think so."

Bob shrugged, and it appeared everyone concurred. Ron crawled over the seat, lifted two hockey straps, and from under the tarp, he lifted out the little grill. He set it on his seat and then returned and grabbed one of the bundles of wood. It was in pieces around two feet long and three inches wide. The boards were different thicknesses, most a half-inch thick, much like the old fence palings. There were newspapers wrapped up with the boards. Ron untied the bundle, took a few pages from the paper, wadded them up, and put them in the grill. He then went back to the spot and grabbed his small axe before stretching the hockey straps back over the tarp. He returned and split one of the boards into small slivers. He placed the slivers over the newspaper and moved to get the beans ready so they could start them as soon as the fire was going.

He went over two seats to where the food was stored and fished out the bag of beans and cast iron pot with a lid. He poured the beans into the pot until it was almost half full. He thought that was close enough to twelve cups. He could measure it next time, if there had to be a next time, and then he paused to hope they would hit land and be safe before there was a next time. He put the lid on the pot and then returned the bag of beans to their place and grabbed the container of water. He filled the rest of the pot with water to just under the lid, put the lid back on the pot, and put the water container back where it was. He then carefully carried the pot full of beans and water over to the grill and set it down.

He took his lighter out of his pocket and started the fire. As soon as it was burning, he put the pot on the top of the grill and grabbed his axe to split another board. When the slivers of wood caught fire, he pushed in a thicker piece. He looked up and realized he was cooking on the seat where they had been rowing.

He said, "Do we want to move this thing on to another seat so we can keep rowing while it cooks? It might not be worth the risk of dropping it."

"No," said Steve, "leave it there. We might be rowing in the wrong direction anyway."

There was some laughter. Ron said, "I hope not, but it will be good to relax. While we can, anyway."

The sea was still calm, and the sun was warm. Ron was keeping the fire stoked while the rest of the crew talked. The children were on top of the tarp on the floor, playing with their little stuffed koala bear. It occurred to Ron there were thirteen people, and while there was plenty of food, he only had two bowls.

He said, "I only have two bowls, so unless anyone else has a bowl, we'll have to take turns eating."

The fire had been going for an hour, and there was still half of the bundle of wood. He had been letting the fire heat up, and he got as much as he could out of it before adding just enough to keep the pot boiling. He lifted the lid and looked in. The beans looked like they were cooked, and surely were soft enough to chew. He thought there was enough firewood for them to have four more fires for cooking. Then he thought he hoped they wouldn't be out there another night, much less four.

He crawled over the next seat and, from under the tarp, pulled out his one-liter plastic salt shaker and two bowls and cups. He asked who had their own bowl and went back to the grill and took the lid off the pot. The children were there first, each with a small plastic bowl. Ron took a cup and dipped it into the beans and then poured into the children's bowls. Then he sprinkled a few shakes of salt over each one. It turned out everyone had their own cup or bowl. Ron dipped in the cup and poured into each of theirs a cup of beans with a few shakes of salt. He was the last, and there were about two cups of beans left, and he declared the extra few beans were the benefit for being the cook. He held the lid of the pot tight over the top and poured the excess water overboard. Then he removed the lid and poured the remainder of the beans into his bowl. The bowl was full. He shook the salt over it a few times and then stirred the beans with his spoon and shook the salt over it a few more times.

He tried a spoonful. Such a simple, plain food tasted so good. He shook more salt over the bowl and ate another spoonful of beans. He declared it a wonderful meal, and there was some mumbling and nodding in agreement. Everyone seemed to be enjoying it, and maybe that was why they were so quiet.

"First one done could start pouring the rest a cup of water," he said.

There were more mumbles and nodding. Julie was the first one finished. She rinsed her and the children's bowls over the side of the boat and then picked up the water container and poured it into a cup. She handed the first cup to her first child. After she drank it, she handed it back to her mother, and then Julie repeated the process for her son and Jin. Sue walked over with two cups, and Julie filled them. Sue handed Charlie one of the cups and drank the other herself. Julie poured herself a cup and drank it and then passed the water container to Mark, who was still eating. When Mark finished eating, he resumed the water-pouring duties until everyone had drunk a cup of water.

There was still daylight. Ron dumped the ashes from the grill overboard and gave the barbeque a rinse before stowing it back under the tarp. He suggested they might as well start rowing again because there was nothing else to do. He stood up and had a look around the other side of the boat. On the horizon, he could see rain coming. It was that hard rain that you can see pelting the water in the distance. He thought how lucky they were to have finished eating.

"Look at this timing," he said."Here comes rain; maybe we should forget about rowing and get under the tarp."

Everyone looked in the direction of the rain and several swore. Ron crawled over the seat and removed the hockey straps and started to unfold the tarp. At least they would have time to tie it down and not get wet. Then a thought struck him, and he called for anyone who needed to move their bowels to do so before they tied the tarp down. Going to the toilet over the side of the boat hadn't been easy for any of them, but it would be impossible during a storm. They had all had to go before then. There was no embarrassment, and when someone would go over the side, everyone else would look the other way. Ron stood up and peed over the side. The boat had started rocking, but it was slow enough that he could manage to stand close to the edge. It was everyone's cue to go, and they did as they watched the rain approach. Ron was the first one finished, and he started unfolding the tarp. They managed to spread it out and tie it down before the rain started.

They were all lying down on the floor or seats and looking at each other as they waited for the rain to hit. The boat started rocking faster, and then it was smacked with a wave. No water came in, and that was a good sign. Then the rain poured. It was deafening. None of them could have heard the other speak. Then Ron got an idea. He picked up his cup, went to the side of the boat, lifted the tarp slightly, stuck his arm with cup outside, filled it with rainwater in a matter of seconds, and then pulled it in and drank it. The rest of them took turns doing the same. As long as they only opened the tarp just enough as they needed to get the cup out, very little rain came in the boat. The rain was a great thing because they were all drinking plenty of water and not having to worry about rationing it.

Ron noticed when someone stuck their arm out there was a stream of water that would come in when they first opened the tarp. He grabbed the water container and set it near the side of the boat just under the tarp. He opened the tarp and held it so the rainwater ran into the container. There was some spillage but not much. It didn't take long before the container was full. He took a sponge he had thrown in the boat with the supplies and soaked up the water from the floor and wrung it outside the boat. It was a handy sponge and only took several soaks before the floor was dry. Then he tied the tarp down again, thinking he should be thankful for the small mercies.

"It might as well be time to sleep again," he said loudly.

Nobody could hear him, though, so he made the motion with hands on the side of his head. He lay down next to Charlie with his back on the floor and his knees bent because there wasn't enough room to stretch them out. Sue crawled over and lay down on the other side of Charlie. The couples all paired up and did their best to get comfortable. The rain kept pouring, and before long, they were all sleeping.

Ron woke and wondered how long he had been asleep. The boat was rocking gently, and the rain had slowed down to a shower. There was no light making it through the tarp. He didn't know if it was night or dark because it was still raining. He opened the tarp just enough to peek out and saw it was pitch black. He thought that even if it was morning, it must be very early, and as long as it was raining, they might as well keep resting. He lay back down and closed his eyes. He had to pee, but he didn't want to open the tarp and stand up. He could hear the others snoring and didn't want to wake them. He tried to sleep and nodded off for a few minutes a few times, but for the better part of a few hours, he just lay there and listened.

He was thinking of what they should do once they were all awake. He could provide breakfast again, but maybe he should save the food until they were all very hungry. They might not make land for another day, or even longer, and they could run out of food. He had seen them all take cups out of their bags, so he thought maybe they might have some food, too. He suddenly thought he couldn't hold his pee much longer and opened the corner of the tarp and stood up. It was raining but not near as hard as he had thought it was judging from the sound under the tarp.

As he peed, he looked around, but he could only see the horizon in all directions. It was light, but the sun wasn't showing. He couldn't see very far because of the rain, and without the sun, he could only guess what direction they were moving. He sat down and was about to tie down the tarp again when Mark asked him to leave it open and then crawled over and stood up in the same spot Ron had been standing. Ron waited for him to finish, and by then, Steve had crawled over and was waiting to take Mark's place. Everyone appeared to be awake. Steve declared he was next and then swapped places with Mark. Ron and Mark crawled to the front of the boat, and the girls crawled to the back of the boat. Mark called for everyone else to line up for the toilet, and they all took turns. Bob and Sue helped Charlie stand up. His ribs were broken, and everyone could see he wasn't feeling too well. The children looked sleepy and unimpressed when they moved to the back of the boat, and after everyone had their turn, they were all sitting quietly.

Ron said, "What do we think we should do today?"

"What are our choices?" said Bob.

"We can row, or we can drift. That's about it."

"Do we know what direction to row?"

"Not unless the sun comes out."

"Then we may as well drift."

Everyone agreed. Ron was going to mention food but thought as long as it was raining and they were under the tarp, it would be a hassle to prepare and eat. He thought about how it could be done, but there was no rush, and he figured they might as well wait to see if the rain would stop.

The children were making motions to their parents and were obviously bored. Jin produced a deck of cards and he and Julie played cards with the children. The couples paired up, and Ron took his place on the floor, on his back with his knees bent, and closed his eyes. He lay there, awake, listening to others natter, and thought he wouldn't be able to sleep. His back was sore, and he tried to relax in hopes that the pain would stop. He could feel the pain from his lower back right up the spine to his neck and under his left shoulder blade. He tried to think of something else. He thought about his son and granddaughters, about what they might be thinking or if they were wondering where he is. Everything would be alright when they were all together. That was the thing to think about now. *Don't be negative. Think of something positive. Forget the pain and go to sleep.* He kept telling himself these things, and a few minutes later, he was sound asleep.

A few hours later, he woke again. The rain was softer, but the boat was rocking more than it had been. His back was worse. His lower back felt like a stiff board, and the pain up the spine to the neck was worse when he straightened his legs. His legs hurt, too, and the muscles just didn't want to straighten. He lifted his knees and then pushed out one leg, retracted it, and pushed out the other a few times. Then he took a deep breath, let it out, and rolled over. He untied the tarp and opened the corner. He pushed himself up with his arms and stood. It took quite a while to stand up straight. The back gave a little at a time; it seemed he would move a fraction and then would need to stop, and then he would have to push it another fraction and stop and repeat several times before he was standing straight. It hurt, but at the same time, it felt good to stand up. The pain sensations running up and down his spine gave him the feeling his back muscles were moving while he was standing still.

The boat rocked, and he nearly fell over the side. He looked around. It was lighter, but he still couldn't see the sun. Other than the sea, he couldn't see anything else either. He was hoping they would see land or a ship or plane or helicopter or anything. He sat back down and looked at the rest of them in the boat. They were all awake. The children were eating raisins, so that answered his question if anyone else had food.

"It's not raining much," he said. "Do we want to get out from under this tarp?"

"Yeah," said Steve, and he started untying the tarp where he was.

The rest of them did the same, and they folded the tarp and placed it back on the floor over the food and supplies. Then they stretched the hockey straps and snapped them on to the hold the tarp and supplies down. Ron again thought about eating, but because they had just gotten everything stowed away, he thought he would bring it up later. The swell picked up considerably, and every few minutes, the boat would fall quickly and smash the water.

Ron said, "Maybe we should aim the front of the boat towards the waves, or we might get tipped out."

"Yeah," said Bob. "Let's do this now."

Steve and Bob took an oar each and aimed the boat at the waves. It was rocking up and down but not crashing and jarring them around like it had been. They all were looking out for the next wave and would hang on when the boat would lift up. It would have them looking up at the sky, and then down the other side, they would be looking at the bottom of the wave. That continued for a few hours before Steve and Bob stopped and Bob called out for someone else to have a turn. Mark and Jin changed seats with them and took an oar each. Ron thought maybe they could see he was struggling with his back and that was why they'd gotten there ahead of him, or maybe they thought it was their turn. Either way, it was a good thing; his back was killing him, and it was trouble enough just hanging on.

Jin and Mark had been using the oars for a few hours when the clouds thinned enough for them see the sun. It was already late afternoon. Ron thought that morning didn't seem so long ago; the day had gone quickly. He had been waiting to mention food, but even if they wanted to cook something, the boat was rocking too much. If they were going to eat anything, it would be the dried fruit.

Ron said, "We can have some more dates and figs. Does anyone else have something to eat?"

The kids put up their hands, and Julie said, "We have raisins."

"We have rice," said Sue, "but we can't cook it in this. We also have carrots."

"We may as well have the fruit," said Ron. "If we are still out here tomorrow night and can't cook, maybe we can all have a raw carrot."

He crawled over the seat and undid the hockey straps to get under the tarp and get the food out. He removed thirteen handfuls of dates and thirteen figs and put them in a bowl. He returned the rest and stretched the hockey straps back over the tarp. Then he went around the boat and gave each person a handful of dates and one fig. He left Bob and Mark for last. He held up one finger in a gesture they should wait a minute while he ate a fig and a handful of dates. Then he handed the bowl to Mark and gestured to Mark and Bob to swap seats with him, which they did. Then he took over the oars while Mark and Bob ate their figs and dates.

Ron rowed for a few hours. He would aim the front of the boat towards each wave and then gain a few strokes down the other side. It was going smooth enough, and he thought they were making good time in spite of the swell, so he wanted to keep going. He was pretty sure they were going the right direction, but whichever direction they were going, it had to be better to move faster. It was getting dark, though, and they would probably be better off under the tarp until morning. Maybe it wouldn't rain, and the tarp wouldn't be necessary, but if they waited too much longer, it would be harder to tie down in the dark.

As he continued rowing, he said, "Maybe we should get out the tarp and hunker down for another night."

"How about we all have a cup of water before we do that?" Steve said.

"Good idea. And anyone who needs to pee or poop, do it now. Even if you don't have too, try."

There was some laughter and chatter as everyone took turns leaking or sitting over the side. Jin and Julie held a child each as the boat rocked up and down. The kids took it in their stride and hugged their parents tight when they were done. Steve and Sue helped Charlie manage. He was struggling with his broken ribs. Mark poured the water as everyone took turns bringing their cups to him. Ron stopped rowing and put the oars away. Then he took a leak over the side and washed his hands.

He dried his hands on his shirt and then crawled over to Mark for a cup of water. He drank his water, and they worked together to get the tarp out. Mark and Bob unfolded the tarp and passed it along.

Ron said, "Let's try and tie it down tight enough so that maybe we can keep the water out even if we do dip into a wave."

They tied down the tarp, and then each of them grabbed what makeshift bedding or cushions they had and made themselves comfortable. This time, instead of the floor, Ron had a bench to himself. He thought that would be a lot better on his back,ashe could put a foot on either side of the bench and stretch out.

The boat rocked most of the night, but the swell had slowed considerably the next morning. Ron woke and opened his eyes. There was light coming through the tarp, so not only had the swell settled down, but it had also stopped raining, and the sun was out. That excited him and made him happy for a second. Then he remembered they were lost at sea and how dire the situation was and thought he might as well keep sleeping if he could. He wondered what had happened to people on land where the tsunami had hit and if they were better off floating in a boat. At least they were alive and, apart from Charlie, were all well. His back was spasming, so maybe he wasn't so well, but it sure could have been worse. He could hear snoring, so he knew someone else was still sleeping, and it would be rude to wake them. They all needed the rest and should use the opportunity to sleep as much as possible. To continue sleeping seemed the smart thing to do, so he closed his eyes again.

He slept for awhile longer. When he next woke, it occurred to him that if a plane did fly over, they would see a boat with a tarp and might not think anyone was in it. He thought he should get up and remove the tarp while they were able and while they could be seen. He opened his eyes. From what he could tell from the light on the tarp, it was a sunny morning and at least a few hours later than the last time he had looked. He rolled over to his side, crawled on his feet and hands to the back end of the boat, and untied the tarp. He peeled it back to let the light in and had a look to see who was awake or still sleeping. Jin and Julie were wide awake, lying with their sleeping children. Everyone else appeared to be sleeping.

Ron rolled up the tarp from the back for around six feet, tied it down, and stood up at the back end where he had opened it. His back was stiff and sore, so he stretched and reached for the sky. He arched his back, rolled his shoulders, and touched his hands over the top of his head. The sun was warm, and the swell calm. The boat was rocking enough that he thought it wise to hold on, and he grabbed the side of the boat as he sat on the back end. At least now, if a plane did go overhead, they might see them. There was no reason to wake anyone else.

He looked out at the sunny, warm day and could only see water in every direction. There was some debris here and there, but nothing close enough to see exactly what it was. He felt the sun on his face, and with his eyes closed, he faced the sun and viewed the red inside his eyelids. It was very relaxing as he thought, *Why get in a hurry to do anything else, and I may as well enjoy the moment while I can.* He sat for a moment and then stood and stretched, moving his neck up and down and side to side, trying to loosen his muscles and get exercise. Everything from his legs to his neck was so tight, and his back was so sore. He felt the urge to move his bowels and undid his pants. He looked into the boat in time to see Jin and Julie smile and look away. He sat over the side and held on tight. Everything went well and was a huge relief. He thought what a blessing that was.

The swell had calmed, and the sun was out, and the morning was so good so far. He thought ahead and figured they could take turns rowing again. If the weather held and the swell stayed calm, they could cook some beans. He thought that if food was going to get scarce, maybe they should eat just once a day. It might be best to eat in the middle of the day. They could row the boat, take a break to eat, and then row the boat until dark and pullover the tarp to sleep again. That is if they didn't find land or have a search plane see them. He thought rescue was unlikely. With the magnitude of the disaster, the emergency services would be so busy for so long that any rescue would be a miracle fluke. It would also be the best thing that could happen to them. He imagined a plane spotting them and a ship or helicopter coming to rescue them at that moment. That would sure be a lot easier than trying to row for land, especially when they weren't sure which direction the nearest land was.

He looked around again and still could see no sign of land. They would have to row away from the sun to where Australia should be, and hopefully, it was not too far. Then he thought that no matter how far it was,all they could do was keep going, so he might as well anticipate it was years away, and then he could be pleased when they found it sooner. He washed his hands and crawled over to the water and poured himself a cup, drank it, and then crawled back to the rear of the boat and stood up. Jin and Julie were right behind him, and the three of them all stood there. They smiled and nodded at each other, and then Julie made the gesture she was about to drop her pants. Ron smiled and looked the other way while Jin held on to her as she sat over the side of the boat. When she was done, she washed her hands in the seawater. Then she helped hang on to Jin while he did his business.

The others started to stir and move around. Ron greeted and welcomed them to the morning and suggested they put away the tarp. They folded it into a smaller square and placed it over the food and supplies on the floor. Then they used hockey straps to hold everything under it. They could be sure their things were tied down and would stay put if there was a swell, but they could only hope it would be watertight.

Ron volunteered to row first, and Jin was standing next to him and held up his hand to join him. They started rowing while the rest of the crowd helped each other do their business over the side of the boat. Julie crawled over to the kids and helped them hang over the side. They all washed their hands and faces in the seawater, and they all drank a cup of fresh water. Every container they had that would hold water, they had filled during the downpour, so they wouldn't run out of water anytime soon. Ron thought about how much water they had and about how many cups of water that was. He thought there must be a few hundred cups, so even if they used six cups to cook a pot of beans, they would have enough for at least ten days, or maybe even two weeks. It seemed even if they didn't touch land before then,it would rain sooner than that, so they might as well drink as much as they wanted and stay hydrated. He thought it certain they would find land inside of two weeks in any case.

Ron and Jin rowed for around an hour. They kept rhythm at a steady pace, and if felt like they were making good time. There was haze on the horizon in the direction they were heading. Ron stopped first, and then Jin. Ron pointed ahead and said it looked like smoke or pollution. Jin nodded. Steve offered Ron and Jin a break from the oars, and Ron agreed. Steve and Mark crawled over and took the oars and the seat, and Jin crawled over to Julie and the kids.

Ron lay down on his back next to Charlie. His back was aching, and the pain was shooting up and down in the way that made him feel like his back was moving even though he was lying still. He closed his eyes and waited for the sensations to stop, which they did after several minutes. It felt better; it was still aching, but it was better, and he could relax. He felt the sun on his face and smiled. He could feel the boat moving and hear Steve and Mark's oars rowing in rhythm and figured they were making good time. At least, it was fast time for a boat without a motor or sail. *Make hay while the sun shines*, he thought. He lay there thinking about the times he had used the saying before and the jobs he'd had over the years. He had a strong work ethic and was proud of it. He let his mind roam over his previous jobs and the past in general while he lay there with his eyes closed and the sun on his face. Before long, he was sound asleep again.

A few hours later, he woke again. He lay there with his eyes open, looking at the sky, and from the sun, he judged it was afternoon. The sun was heating up, too. There was only an intermittent breeze and a light swell. He looked around the boat and noticed Bob and Mary had taken over the oars. As he noticed everyone else, they noticed him, too, as they exchanged glances and smiles. Even Charlie managed a smile, but nobody said a word. Ron's back was still as sore as ever. He rolled over to push himself up. It took quite a while to straighten his back, and he did well to keep his balance in the process. He did his stretches and went through what was becoming his technique while in a boat: reach for the sky, stick his arms straight out to the side and make little circles, hold his arms straight out and touch his hands over his head, bend his neck and look up and to each side, arch his back, and shake the hands and arms out at the end.

When he was done, he was still sore, but at least he could move about and straighten and bend his back and joints. He was looking around all sides of the boat while he did his exercise and noticed more debris in the water, including a large piece of something floating ahead of them. He pointed ahead and motioned to Bob and Mary to have a look, too. Mark and Steve stopped rowing and looked to see what was ahead.

"What is it?" asked Bob.

"I don't know," said Ron, "but it looks like a frame of some kind. If there's any timber in it, we should grab it and use the wood for a fire. We could cook some beans and save the wood we have."

Bob and Mary swapped seats with Mark and Steve, and the rowing started again. They only had to change direction slightly, and five minutes later, they were ten or twenty meters away from it. It looked like part of a wall with a timber frame and Styrofoam insulation. Maybe it was once part of a cool room or something off a fishing boat. When they were close enough, Deb and Sue, who were at the front of the boat, grabbed it and held it up. It was about two meters by one meter wide, with thin plywood and a frame made from two by fours. The Styrofoam appeared to be stuck to the ply.

Ron said, "I don't know what it is, but it is a gift. We'll use the timber for firewood."

The girls passed it on to Jin, who passed it to Mark, who passed it to Ron, and he looked at it. The Styrofoam hadn't been glued; it must have been poured in or something to have caused it to seep into the gaps of the frame, and that had held it in place. Ron thought it was lucky it had, or it wouldn't have floated. He grabbed the end of one of the two by fours and, with the other hand, the edge of the ply and pulled it apart. It didn't take much effort, and the Styrofoam just dropped out. Ron commented they were lucky the thing hadn't come apart and sunk before they'd found it. He punched the ply and banged the timber on the boat seat to separate them. In little time, it was in pieces. There was one two by four about two meters long, there were three that were about a meter long, and there were three more about a foot long. He thought about throwing the Styrofoam over the side and then thought better.

He said, "We can use this Styrofoam for a pillow, or at least some padding under the head to sleep on."

The Styrofoam had broken into four pieces worth saving, around half a meter by a foot and four inches thick. Bob took two pieces and handed one to Mary before taking the other, placing it on his seat, and then sitting back down on it. Mary did the same, and they started rowing again. Ron crawled over to dig out the portable barbeque. He undid the hockey straps, picked up the barbeque, and put it on the seat. He located the big pot and his bag of beans and placed them next to the barbeque. Then he crawled over to pick up the timber and bring it back.

He said, "Does anyone want to fix up these beans and the pot while I start the fire?"

"Yeah," said Deb, "let me do it."

"Make it thirteen cups and then a half for good measure."

."We have some rice," said Julie, "if you want to put it in."

"Or we could have it tomorrow night," said Ron. "There's enough wood here for another fire."

"Is that all the food we have?" asked Steve.

"No," replied Ron. "I have some rice, and there are enough beans for a few more feeds. We should find land sooner than we run out of food, but it may be wise to be careful."

"You can never be too careful," said Deb.

Ron nodded. "Yeah, I've heard that saying before. Let's vote on it. Everyone who wants rice and beans tonight instead of beans tonight and rice tomorrow night, raise your hands."

"We don't know what the weather will be tomorrow," said Bob. "We may not be able to cook."

"Yeah," said Ron. "Well, think about it, and we'll vote."

The vote went for both rice and beans that night, and Mark said, "Make hay while the sun shines."

"Yeah," said Ron, "I've heard that saying before, too, but I'm not sure that's what it means tonight. Ok, no problem, rice and beans tonight. We'll live it up while we can." There was some laughter and light cheering, and he added, "We're doing alright considering the predicament we're in."

He started the fire while Deb put the beans and rice into a pot and covered them with water. As soon as the fire started, they put the pot on top of the barbeque. Ron asked if they should start rowing again. Sue and Mary took over the oars and called out, "Girl power!" as they managed a solid rhythm. They kept going strong for several minutes and then tired and stopped.

Mary said, "How about slow and steady wins the race?"

That got a laugh from most everyone. They kept a slow, steady pace while Ron kept feeding the fire only as much as he needed to. From time to time, he removed burnt ash carefully so as not to waste any of the heat. He was using shredded slivers of the thin plywood. He split one of the two by fours into several shreds and the other into four. That burned for half an hour, and he asked Deb to add the rice. She did while Ron split another two by four. Another twenty minutes and the rice and beans were cooked. Ron looked at the wood and said they hadn't used anywhere near half of the wood. He declared it time to eat, and Mary and Sue stopped rowing, and everyone located their bowl or cup from wherever they had it stowed.

Deb served the beans and rice using her cup to dip into the pot and then pour into the next waiting cup. Someone would crawl over to have her serve them and then crawl back and sit down before someone else would get up. The children went first. Deb shook a generous amount of salt over the top of each serving and sent them on.

She said, "First, I'll give everyone a cup each, and then we will see what's left."

Everyone lined up with their own dish or cup, and Deb was very careful not to spill any as she served them. After everyone was served, everything was silent while they ate. Deb let them know the pot was still half full, and there were groans of approval. As soon as Deb finished her cup, the rest of them took turns crawling over with their bowl to have Deb serve them again. She served each of them a little more than half a cup. After she had served everyone else, there was about a full cup for her, and she declared everything had worked out pretty good. When they were all done eating, they rinsed their bowls or cups in the seawater over the side of the boat. Ron cleaned up the barbeque while Deb rinsed the pot. Ron put it all where it was before and snapped on the hockey straps to hold it down.

"I feel good," he said."Does anyone want to join me with an oar?"

He actually didn't feel so good. His back was still aching, but he thought getting active with the oars might help him keep his mind off the pain or loosen up his stiff muscles somehow. Mark volunteered, and they started rowing at pace, not too fast so as to tire themselves out, but about as fast as they could otherwise. The conditions were still calm, and there was only a light breeze and not a cloud in the sky. Ron was sore but feeling strong. He thought they were all in pretty good spirits, and now they had been well fed.

He said, "This is just like a day out boating with a healthy meal and exercise. People pay heaps to have vacations like these."

"I'm going somewhere else next year," said Mark.

"We really are lucky, though, that we survived everything so far and are in reasonable shape."

"Charlie's not so good, though."

"I know, but he should survive ok. All in all, we're doing better than most of the world. All we need to do is find land and get out of this boat. Imagine all the other places that have been flattened by earthquakes or wiped out after a volcano, or the radiation like in Korea and what's left of Japan."

"I don't think anything is left of Japan."

"Well, that's what I mean. We are lucky. Though I suppose radioactive water or whatever it is will reach us sooner or later."He paused and looked around before he said, "It's funny how you wouldn't know the trouble the world is in right now if you were just looking at what a beautiful day it is. Make hay while the sun shines, or just be happy and row your boat while it shines."He started singing "Row, Row, Row Your Boat."

When he started the second verse, Mark joined in, and before long, they were all singing, laughing, and doing the harmonies. They took turns coming in after the first verse. They sang for what seemed like ages, and whenever they stopped, Ron would say, "Second verse, same as the first," and they would start again. Everyone was smiling, and the kids were stomping and clapping.

Ron and Mark rowed while they sang, first "Row, Row, Row Your Boat" and then "A Hundred Bottles of Beer on the Wall." After that, they rowed for a few more hours. Then Jin and Julie crawled over and offered to take over the oars. Ron smiled, stood up, and stretched. Mark crawled over the next seat and sat down next to Amy. The kids came with Jin and Julie and sat down on the seat in between them. As soon as they started to row, they began singing, "Row, Row, Row Your Boat" again with gusto. Their oars were in rhythm, and the kids were clapping their hands. Everyone started laughing, and by the time they reached the second repeat, they were all singing again.

Ron's back was killing him, so he stood and stomped his foot with the beat. He changed feet with every verse. That was to keep his exercise even. He thought the exercise would help his back, and it was fun. Plus, he wouldn't let on that his back was hurting. They would think he was just having a good time. He added his arms in a rowing motion for one verse and then arched his back and straightened up, lifting his hands as if he were raising the sky. The kids loved it and were laughing.

When a slightly bigger wave made him lose his balance and fall, they laughed even harder. Ron grabbed the side of the boat, and when he had his balance, another wave hit. He let go and sat on the bottom of the boat, and then he lay on his back with his knees bent. He had exaggerated the fall somewhat for comedic effect, which also had the kids cracking up. Then he started kicking his knees up in the air in rhythm with the next verse. The kids loved that, too. He stayed with his back flat and used his legs or arms to do something different for each repeat. After several more repeats, he was just clapping his hands. The singing died down, and then the kids started it going again. Ron lay there on his back with his knees bent and hands on his chest and tried to sleep. As he was dozing off, he heard one of the kids call out for a hundred bottles of beers on the wall, and two or three voices answered no. Then they started another repeat of "Row, Row, Row Your Boat."

He thought maybe they were all going crazy, but the singing was keeping their spirits high, and that should provide a better outcome. It seemed to make sense that if their spirits were low and they were bawling, that would provide a worse outcome. If their spirits made no difference to the outcome, at least they had spent that day happily singing and remaining positive instead of crying with despair.

Chapter Four
Dark Nights and Days

Ron woke and opened his eyes to darkness. The boat was rocking, and he could hear snoring. Again, it took him awhile to remember where he was, and then he knew they must have let him sleep and tied down the tarp for the night. He must have really been tired to sleep through that. They had placed a piece of the Styrofoam under his head because he was the one who said they should keep it and use it.

He lay there thinking about the good spirits they'd been in before he'd fallen asleep. All they had been through and what was happening around the world was dreadful. Still, they were having a reasonably good time. That had to be a positive thing and better than crying or giving up. He thought about what might have happened after the last tsunami and what they might see when they did find land. He was hoping his son and family would be safe. He knew they must be worried about him and probably thought he'd drowned. He thought maybe he just hadn't drowned yet. Then he thought he couldn't. They couldn't have gone through everything they had with all the luck plus the huge effort and then die. They must be meant to survive. Maybe there was some grand plan or something or someone watching over them. He decided that whether there was or not, they could never know, and the best thing to do was stay positive. They would just keep doing what they were for as long as it took. Stay happy and never doubt they would survive.

He closed his eyes and drifted in and out of sleep until the morning. When he decided to wake, he opened his eyes again and could see it was lighter but still very early. It was the sound of Julie helping little Eddie have a dump over the back of the boat that woke him. She had opened up the back corner of the tarp like he had the morning before. That had let the light in, and Ron thought that an opportunity to untie the tarp to where he was and he could have a sit over the side of the boat, too. He crawled over and untied the ropes on either side of the boat and rolled the tarp back. The swell was calmer than he had ever seen it. The sun hadn't broken the horizon, but there was enough light to see it was partly cloudy and very still.

As he stood up straight, his back made a noise and hurt enough to remind him he was going to be miserable again. He would be until they found land and he was able to stretch out and really flatten his back for as long as he needed. He relieved himself over the side of the boat and then did a series of stretches before assuming the squat position over the side with one hand hanging on to the part of the tarp that was still tied down. When he was done, he rinsed his hands in the seawater. The others had woken up and opened the tarp at the front of the boat, too, leaving the middle covered.

They thought it a clever idea to keep the boat partially covered with the tarp so they would have some shade. That way, Charlie could be in the shade, and there was enough room for a couple of people to take turns sleeping or just being under there with him. Charlie was in a lot of pain and taking it well, but he needed help to move or do anything. The others would feel good about themselves when they helped him with a drink or held on to him when he leaned over the side of the boat. Looking after Charlie was another thing boosting their morale, and his welfare was important to them all. Charlie would show gratitude with a smile and a thank you, and that also boosted their morale.

They all had two cups of water, and Ron put away the container and made sure the hockey straps would hold in place. It wasn't much of a risk while the conditions were so calm, but you never knew when a freak wave would hit or the wind would pick up. He looked out over the horizon, and the sun began to show itself fully. He was still uncertain which way to go, but he figured if he had been on the beach fishing and the sun had come up, he would be looking straight at it. So, with that logic, it made sense to row directly away from it.

Ron asked to row first, and Steve joined him, and with an oar each, they glided through the calm water in a steady rhythm. They rowed for around four hours, stopping only a few times for a few minutes. The sun was higher and heating up. The shadows were shorter, and Ron was even less certain they were going in the right direction. He thought it didn't really matter. It was just that they needed to be doing something to help them find land and also get exercise and help the day go by. Maybe it could help, and it couldn't hurt, so they should keep going the way they were. Ron asked who else wanted a turn rowing, and Bob and Mark took over the oars while Ron and Steve lay on the floor under the tarp with Charlie and Sue.

Bob and Mark rowed for a few hours and stopped. It was afternoon, and the sun was high and bearing down, and Ron suggested they should take a break and cook some beans while the sun was so hot. There were sounds of agreement and none in dissent, so he undid the hockey straps and dug the barbeque out. He set it up and split some kindling to start the fire. Meanwhile, Deb and Sue doled out thirteen cups of beans, dropped them in the pot, and covered them with water. As soon as the fire started, they put the pot on top of the grill. Ron kept the fire going, and again, he was careful to burn only as much timber as necessary. When the beans had cooked, everyone took turns to crawl or walk over with their cups or bowls, and Deb gave everyone a cup each. They had the cooking down to a routine.

Ron noticed they were all smiling and happy again, and he guessed the food had something to do with it. When they finished eating, they washed up and stowed the barbeque and pot. Jin and Julie volunteered to row. Eddie and Jesse joined them and started singing "Row, Row, Row Your Boat" again. They all started singing again, rather quietly at first, but they gained volume gradually. Soon, they were all smiling and singing, and it was very much like the day before. After a few dozen repeats, they took turns leading any song they could remember. One would have a turn and think of a song and then sing it out loudly, and any of them who knew the words would sing along. The singing eventually stopped, but their spirits remained high. Jin and Julie rowed for a few more hours, and then Mary and Amy took the oars for a few hours.

The sun was getting lower when they noticed clouds straight ahead. Then they could see the rain hitting the water in the distance.

Ron said, "I don't know if that's a good or bad sign. Looking at this, the shore must still be a long way away."

Bob and Mark covered the rest of the boat with the tarp. They managed to all be under it and had it tied down when the rain hit. There was some wind when it first hit, and a little bit of movement in the swell, but then they could tell the rain was falling straight down. The swell was still pretty calm, but the rain was teeming. Ron slipped his cup out of the side of the tarp and filled it. He brought it in carefully and drank it. Everyone did the same, and then they all made sure they had another cup before they tied the tarp tight again. Ron started singing the song "Pennies from Heaven," and some of the other joined in. When they finished, Bob started singing "Raindrops Keep Falling on My Head," and the others joined in. They took turns singing songs about the rain, or at least the parts of the songs they could remember. They sang about stormy weather and riders on the storm and rain in September. A lot of the songs everyone knew a few lines to, but nobody knew all the words to all the songs. They sang and laughed for around an hour. They all had another cup of water and then got comfortable to sleep for the night.

The rain would pelt down, then let up for awhile, and then pelt down again. The sound of the rain was loud but peaceful and an asset to their sleep. Sometime in the middle of the night, little Eddie had to pee, and when the rain let up, Jin and Julie untied the tarp at the back end of the boat. Ron woke and saw what was going on and called out to leave the tarp open so he could go next. As it turned out, almost everyone had to pee and took a turn. It was difficult for Charlie, and he needed two people to help him. Even though he was struggling, it was obvious to everyone he was improving. Ron's back was still killing him, but he thought he was at least better off than Charlie. They all got back to their places and were drifting off to sleep again. Then the rain started teeming again. Ron thought that was perfect; the rain had let up to let them all pee and then poured again to help them all get back to sleep.

Ron woke again, and it was morning. It was still raining, but there was some light getting through. The swell was rocking the boat and moving it about, but it was still comfortable compared to what it had been. His back was sore, and he had slept all he could, so he crawled to the back of the boat and untied the tarp to roll it back just enough so he could stand straight. The rain was his in his face, and with the pain in his back when he stood, he was feeling somewhere between delighted and in excruciating pain. On the one hand, it hurt like crazy, but on the other hand, he knew it was the best thing for him. His back eventually straightened one vertebra at a time. He held on to the edge of the boat, arched his back, and let the rain hit his face. His shirt was drenched, so he took it off and wrung it out. Then he did it again. He thought he might as well wash his clothes while he was out there, and since his jeans were wet, he took them off, too. He was careful not to lose the things in his pockets as he wrung them out.

He noticed the water was welling up on the floor, so he mopped it up with his shirt until it was reasonably dry. Then he drenched the shirt in fresh rain and wrung it out again. He threw the jeans and shirt to the spot he had been sleeping. He closed the tarp for a minute and then grabbed his cup and went back out. After standing and stretching for the sky as high as he could, he filled his cup, drank the water, and filled it again. His body needed the stretch-to-the-sky exercise for his back as much as it needed the water. He didn't want to let on his back was killing him, and it was one way to get much-needed exercise without drawing attention to the fact his back was spasming. He stood in the rain for a few minutes, rubbing his body with his hands while stretching and reaching. He had a pee and washed his hands before coming back under the tarp.

As he was about to tie the tarp back up, he saw Julie, Jin, and the children crawling towards him. Julie had a wide, square metal bucket that looked like something someone might soak their feet in. Ron thought what a great idea that was. They could fill it, bring it in, and have a wash. Jin motioned to Ron to leave the tarp open. Ron smiled and then crawled under the tarp towards the front of the boat to give them more room at the back.

"What's going on?" asked Steve.

"I just had a wash," said Ron. "Everyone should get in on that while we can. We can take turns getting to the back of the boat and then have a wash and mop up the floor when we're done. It's raining just enough to have a good shower but not so much to flood the boat. You could say, 'Make hay while the sun shines and have a shower when it rains.' We should fill the water containers while we can, too."

Ron thought about filling the water container at that moment. He looked to the back of the boat and saw Ed and Julie were naked. Mostly, he noticed Julie was naked. He could only see their bottom halves, as the rest of their bodies were above the tarp, so they couldn't see him staring at Julie either. He didn't want to embarrass them, so he thought it better to leave the water container for later and remained under the tarp at the back of the boat. He gawked at her petite and perfect bottom and then looked away in case someone noticed him.

He thought about how superb Julie was and how long he had been alone. He would never interfere with a marriage, but if anything were to happen to Jin, he would like to look after Julie. Sue, too, he thought. The others not so much. He would go with any woman that would have him, but he would not want an unattractive woman as much as he would want Julie or Sue. They were his idea of beautiful: petite, feminine types with beautiful eyes and smiles. Most of all, it was their thin but muscular legs and buttocks. Then he was overcome with guilt. How could he be thinking of such of a thing at a time like this? He wasn't going to do anything about it or hope something would happen. He was going to help Charlie get to a hospital, and then they were all going to go their own happy ways, and they were all going to be ok. All they had to do was find Australia. It was such a big place, so how hard could it be?

After Jin, Julie, and the kids had a wash, they filled the square bucket and carried it over to Charlie and Sue. Sue thanked them and started washing Charlie with a facecloth Julie had given her with the bucket. In the meantime, Steve and Deb went to the back of the boat and had a wash, and so did everyone else. Ron stole a look at Sue as she bathed. He thought how magnificent she looked, too, and then he felt the guilt again. She came back from her wash topless and wearing only little panties. He looked and then looked away and then looked again as she helped Charlie with a cup of water.

When everyone had washed, Ron took the water container to the back of the boat, opened the tarp, and made a crease so the rainwater funneled into the container. He took his cup and bailed out what water he could. Then he filled his cup with clean water and drank it. He tied down the tarp and lay down on the seat.

He said, "Anyone who wants to use the facilities, come on back here. You won't bother me much, and less water will get in the boat back here."

He flattened his back on the seat with a foot on either side and put a piece of Styrofoam under the back of his head. He managed some sleep, but for most of the time, he just lay there and let his mind drift. He thought that was much like the boat being adrift. He thought about his past, and he thought about what his new life in Canberra might be like. He thought about what the others would do once they found land. Would they all be around after that or never see each other again? He wondered about his granddaughters and what they must have been thinking about what was going on. He wondered exactly what was going on. There were all the earthquakes and volcanoes before, and now the latest tsunami. He didn't know what had happened since the tsunami had hit Australia. The way things had been going, there might even have been another tragic event since then. It could be they would find more tragedy when they found land, but surely, it would be better than living on a boat lost at sea. Then he thought maybe it could be just as awful in some other way. What would the future look like in twenty years? He would probably be dead, but his granddaughters would be adults who might have children of their own. What would their lives be like? Would they be comfortable or have to struggle to survive? His mind wandered until he fell asleep.

He woke several times to the sound of someone crawling to the back of the boat to get out from under the tarp. Then rain hardly let up, and when they opened the tarp, he would feel the rain on his right side. When he needed to pee, he didn't have to go far. He just rolled over and crouched as he untied the tarp and then stood up. It was the same thing again with his back, which was hurting like crazy, while he kept trying to think of something else to keep his mind off it. It seemed such an effort and took so long to stand straight that he was soaked by the time he did.

He noticed it was nighttime, so that was a whole day they had drifted in some undetermined direction. They would drift again all night, and maybe in the morning, the rain would break. He took his time and stretched his neck while he peed. He moved his head from side to side and looked over each shoulder. The rain was falling straight down and wasn't too heavy. It wasn't too cold either, so he figured he would be able to lie down like he had been and his shirt would eventually dry. Maybe he would wake up in the morning and feel great. Maybe they would all wake up in the morning and see land right in front of them.

Ron woke early the next morning. It hadn't been easy trying to sleep most of the day before and then at night, too. He could hear the rain had slowed down, and the boat was hardly moving. It was still dark, but he felt sure it was morning. He had slept and woken up four or five times since he'd last stood, and the last time, it had felt like he had been deep asleep for a long while. The pain was across his entire back, from the shoulders to the hips. It felt like his back was a piece of plywood, and his legs were stiff, too, but the worst pain was right up the spine. He thought some of it was from the hard surface, but for the most part, it was because he hadn't been moving. He rolled over and nearly passed out, causing his vision to blur into tiny spots. When he found his balance, he crouched, untied the tarp, and stood up. It was the same thing as before with his back, only much worse.

At least the rain had slowed to a drizzle, and it actually felt good on his face. It was dark, but there was enough light to know it was morning. He wanted to scream in pain, but he remained silent as he reached for the sky and tried to stretch his spine and make it as straight as possible. He stretched for several minutes while looking around the boat and at the sky. It was calm, and the rain was light, so that was a good thing. It was much better than a huge swell or a storm.

The sunny day they'd had before had been pretty good; it hadn't been unbearably hot, and another day like that would be great. If they caught another sunny day and the conditions stayed as calm as it was, he thought it would be a good idea to have a swim. He could tread water for ten minutes or so. That was always good for his back. He couldn't tread water much longer than that without running out of air, but he could hang on to the side of the boat and kick along. He could get some exercise and get the weight off his spine at the same time. It would also make more room in the boat, and maybe one of the others could stretch out for a proper sleep. It was such a good idea that he could hardly wait to tell the others, but he would wait a few hours and see if the sun would come out. The rain was even lighter now, like a mist, so he thought it would be okay to mop up later, and it would be helpful to his back to stand there for as long as he could.

He stood there for more than an hour, stretching and shifting his weight. When he stood still for too long, the stiffness would return, and the pain would worsen, so he kept moving his arms and shoulders. He would lift one foot and then the other, and at the end of every new exercise, he would arch his back and then straighten as he reached for the sky. The rain stopped, and the sun showed itself for a moment, and then the mist returned, and it clouded over again, all before he finished.

Mark crawled to the back of the boat and untied another section of the tarp. Then he stood up and relieved himself. It had stopped raining completely, and the clouds were thinning out. Everyone appeared to be awake but not moving much.

Bob called out, "What are we doing today?"

"Let's get this tarp up, and I'm going for a swim," replied Ron.

There were several responses in unison: "What?"

"Yeah," said Ron, "I figure I can tie a rope around me and go over the back. I can hang on and kick to help push the boat along. It looks a good day for a swim, and we may as well all have a go while we can. If two of us are swimming, two are rowing, two are sitting with them, and four are in the next seat, we can fold up the tarp, and three can stretch out and use it as a bed." There was silence, and he said, "What do you guys say; do you want to try that?"

There was enough consent to consider the motion carried. They untied the tarp, and Ron used a piece of that rope to tie a loop around his body and under his arms, and the other end, he tied to a ring at the back of the boat. He sat at the back of the boat with his back to the sea, pushed off, and dropped into the water. He sorted out the rope, and it was about six feet long. There was nobody rowing, so there was no need to start kicking. He treaded water and felt almost immediate relief in his back. The pain was shooting up and down both sides under his shoulder blades, and there were small areas of muscles cramping and tightening and then letting go. It was pain, but it felt good at the same time. When he started to gasp for air, he used the rope to pull himself to the boat and rested. A few minutes later, he started kicking.

Jin and Julie and the children sat on the middle seat and started eating raisins. Bob and Mark took the oars, and Mary and Amy sat with them while Steve, Deb, and Sue folded the tarp to make a bed at the front of the boat. They helped Charlie get up and crawl over to get on the bed, and then Sue and Deb lay down with him. Steve crawled to the back of the boat and announced he was ready for a swim. He tied a rope around himself and the other end to the other ring on the back of the boat and dove in. Bob and Mark waited until everyone was settled and then started rowing with a slow and steady rhythm while Steve and Ron were in the water hanging onto the back of the boat and kicking to push it along.

They pushed along for an hour or so, and Ron called for a rest. They stopped, and he and Steve had huge smiles on their faces. Ron was gasping for air but obviously enjoying the water. The pain in his back was gone, and now his lung worried him, but apart from being out of breath, he was feeling great. The water was still calm, and the sun was getting through in patches. Steve was enjoying it, too. Bob and Mark were sweating, but they smiled when they saw Ron and Steve smiling. One of the girls called out and asked if they wanted raisins. Julie was moving around with a container and pouring into everyone's hands a good dollop of raisins. Ron asked if she could drop some in his cup and he would have them when he was back in the boat, and Steve asked for the same.

After Bob and Mark ate their raisins, they started rowing again, and Steve and Ron started kicking again. They moved along at a steady pace for another hour or more.

"I've had enough," said Steve. "Let's stop."

Ron was breathing a little easier than the first time they stopped, and he said, "What's the matter? Are you too tired already?"

"No, I'm just thinking of those raisins."

Ron laughed. "Me. too, but I'm going to build up the hunger for as long as I can take it."

Steve started climbing into the boat, and he called for help. Bob and Mark grabbed an arm each and hoisted him upwards and into the boat. Steve untied himself, and Bob grabbed the rope and tied it around himself.

Bob said, "A swim looks like a bloody good idea," and he jumped in.

The sun had won the day, and it was heating up. They were enjoying the swim, but even on a warm day, the water can be cold, and eventually, Ron started to shiver.

He said to Mark, "Are you anxious for a swim?"

"No, I can wait," Mark replied.

"Somebody help me get out."

Steve and Mark helped Ron climb into the boat. He went straight for his raisins and started eating. They were wonderful, and he felt great. He thought, considering their predicament, that things were great at that moment. All they had to do was find land soon, and then everything would be hunky dory. They all went swimming in turn. They thought that if everybody jumped in at once, they might not be able to climb back into the boat.

Jin and Mark picked up Charlie and helped him slide over the side of the boat. He was in a lot of pain when he first went in, but they could all see the expression on his face change after a few minutes. He was smiling. With one hand on the boat, he undressed and dropped his clothes onto the boat's floor. He called out for them to give him plenty of room, and that started the laughter. Eddie and Jessie were in the water, splashing and kicking and showing off that they knew how to swim. Everyone had a comment about how good it felt to get in the water. Ron said it was because they were taking the weight off their backs. He didn't want to complain about his back, but he hoped they were seeing the pain he was in and knew he wanted to get back into the water to ease it.

He said, "I'm ready to get back in and start kicking again unless anyone else wants to go next."

He jumped in the water and grabbed on to the back of the boat and started kicking. Mary and Deb took a turn with the oars. In a few strokes, they had their rhythm and were pushing along at a steady pace. Every so often, a bird would fly past. It would fly over and have a look at the boat and then keep flying. They always flew in the direction they were paddling, so everyone concluded they were going in the right direction. The fact they were seeing birds must mean they were getting closer to land, or so they thought.

They pushed ahead for another hour, and then Sue called out she wanted another swim. Ron watched her fix the rope and get ready with a huge smile on her face. He thought about how beautiful she was and how much he liked being around her, and then he reminded himself how much he liked Charlie, too, and admonished himself for thinking about her. He looked at the sky instead. It was still sunny, but there were clouds in the distance.

Mary and Amy took a turn with the oars while Steve and Deb took a turn stretching out on the tarp and had a nap. They pushed ahead for another hour or so before Ron asked to stop. His breathing wasn't as bad as it was the first time they'd stopped, but he felt he'd had enough time in the water for one day. Bob and Mark helped him into the boat. His fingers and toes were white and wrinkled, but his back pain was gone. He was tired, but he felt so much better.

"Are we going to have beans today?"asked Mary.

"What, eat twice in one day?" said Ron. "We're living like kings."

There was some timber left from what they had picked out of the ocean, but Ron had to dig out and use some of the original bits of firewood he had packed. It was well into the afternoon when they started the fire, and the sun was casting a long shadow by the time the beans had cooked. When they finished eating, the sun was setting. There were fish coming up to the surface from time to time, and suddenly, a swordfish broke the surface and showed himself for one brief moment. Ron looked out at the calm and view and thought once again they were blessed. As bad as things were for them, things were pretty good, too.

"Should we get the tarp over us before it gets dark?"asked Mary.

"Yeah, we better," replied Ron. "We can't count on it staying this calm all night."

Ron was a little disappointed he hadn't gotten a turn to stretch out on top of the tarp, he was but happy to have had a spell swimming in the ocean that had fixed up his back. He rinsed off and cleaned up the barbeque with seawater over the side of the boat. He packed it away and made sure the hockey straps were holding it down as tight as possible. They all had another cup of water and then rinsed their utensils and relieved themselves before setting up a spot to sleep. They stretched the tarp over and tied it down, and Ron remained at the back of the boat and said they should leave the last bit open for the time being and he would stand there for awhile and tie it down before he went to sleep.

He didn't have much room for sleeping, just enough to flatten his back but not enough to stretch out his legs. He thought that the longer he remained standing, the less time he would be in the one position overnight. Not moving, especially while sitting, would always cause the stiffness to return, but in the boat, it seemed much worse and more difficult to stay limber. He stood, stretched, and moved about in place, lifting a knee, swinging a foot, and then changing sides. Then he would rest a few moments and start again. His mind slowly considered food morale and various plans of action for the next day while he kept his body busy. He stretched and moved everything from his toes to his head and made little circles with his arms to the sides and above. He moved about for a few hours before relieving himself and tying down the tarp.

He found his place to sleep and was so tired after all the exercise and what had happened during the day that he was snoring in short time. He woke several hours later and noticed his back was stiff again, but it was much less painful than the night before. He opened his eyes wide, but it was completely dark. The boat was dipping up and down, and every once in awhile it felt like it was being slapped by a wave. He hoped it was moving in the right direction and thought how great it would be if they woke up and could see the beach. He thought the boat felt like a rocking cradle, and he couldn't see a thing, so he might as well enjoy some sleep. Then, once he was rested, maybe he could wake up and see land. If they didn't, they would just start rowing again. Either way, the best thing he could do at that moment was sleep, so he closed his eyes, and within minutes, he was sleeping again.

Four or five hours later, he woke again. This time, he could see light. He rolled over and crouched as he untied the tarp and then stood up. The pain in his back returned at that moment. He had stood up too fast, he told himself. He should have stood up slower, a little at a time. He stopped himself from groaning out loud before he managed to stand up straight. He didn't want anyone to know how much pain he was in. They all had enough to worry about. He stretched his arms and reached for the sky. At the same time, he slowed his breathing and told himself to relax. He tried his best to stay reasonably still while he waited for the pain to stop. He eyed the sky and noticed it was cloudy, but it didn't look like rain. The tarp was okay the way it was for the time being, and as soon as the others woke up, they could lift it and start moving again. In the meantime, he would stand there stretching and keep his mouth shut and his back straight.

By the time everyone was awake, the sun was up, and the clouds had thinned out to reveal a bright day.

Ron said, "I still have enough figs for us all to have one, and there are still some dates. Maybe we should think about rationing what we have left in case we have to spend a few more days out here. There is only enough wood for two more fires, and then we won't be able to cook the beans and rice."

They discussed various options and decided they would eat the dates and figs in the middle of the day and save the beans for the next day. They couldn't be certain the conditions would hold to enable them to build a fire the next day. If that was the case, they would likely be out at sea until conditions improved, so they would eat beans the next day if conditions allowed.

Ron volunteered to swim first, and he staked a claim to lying down on the tarp in the afternoon. He and Sue were in the water first. Bob and Mary took the oars. After an hour, Sue called for a break and traded places with Deb, and after another hour, Deb called for a break, and Steve took her place in the water. Ron was enjoying the water because it was helping his back. Jin and Julie took a turn with the oars. Ron and Steve swam for a few hours before Ron called for a break. He climbed back into the boat and dug out his food.

They all ate a fig and a date or two, and then Ron crawled onto the tarp and stretched out next to Charlie. It was the first time since they had jumped into the boat that he could flatten his back and stretch his legs straight out. He did a few leg stretches and then lay still and enjoyed the moment. Mark and Amy took a turn with the oars while Jin and Julie, with the kids, took a turn in the water. Ron closed his eyes, and within a minute, he was sound asleep.

He woke and went back to sleep a few times when, every hour or so, the group stopped and changed places with the oars and took turns swimming. When they were rowing, the oars made a squeaking noise in rhythm, and when they stopped, the absence of noise seemed to wake him. When they started rowing again, he would fall asleep. After four or five stops, it occurred to him the sound of the oars was like the sound of rain on a tin roof or a lullaby. When he woke again, he thought he had enjoyed a sound sleep and it was time to give someone else a turn to stretch out, so he rolled over and stood up slowly. It was much easier than the time before, and that made him smile. He felt good and asked to take the oars. Bob and Mary were rowing, and Mary said he could take her spot.

Ron and Bob took over the oars and found a comfortable rhythm in short time. When nobody was in the water,they were somewhat crowded, allowing for the space on topof the tarp. Charlie and Sue were on the tarp, and Mary joined them. Ron and Bob kept the slow, steady rhythm with the oars for hours. The shadows were getting longer, and they figured they had three or four hours of light before they would need to stop and get set for the night. Ron decided it would be fun, and maybe helpful, to pick up the pace before they stopped rowing, and he called for Bob to help him go faster.

Ron started chanting, "Heave-ho, heave-ho."

Bob answered, singing, "Row your boat," and so it went in call and answer style.

Then the kids joined in with "Row your boat," and Jin with "Heave-ho."

Every new repeat, someone else would join in until they were all chanting, singing, and clapping their hands. It was smiles and laughter, and the pace gradually gained to a new high when they were cutting through the water at a speed that amazed them all. After twenty minutes or so, the chanting and singing stopped, but the pace remained the same for another twenty minutes. They eventually slowed down to the steady pace they'd had before the singing had started. The exercise had lifted their spirits, and they felt strong and relaxed as they continued. About an hour later, they could see the sky ahead changing. It was very dark, and there were flashes of lightning inside the clouds.

"Uh oh," said Bob. "Everybody, let's get this tarp over us."

They quickly scrambled and got to work, and when the tarp was rolled out, they tied it down tight from the front to the back. They were about to cover the boat completely when Ron stopped them and declared it was time to pee or poop while they had the chance. He stood up at the back of the boat and relieved himself while others did the same. When everyone had finished, the wind was suddenly stronger. They had to struggle to hold on to the tarp and tie the last knot tightly. The wind was wild, and only seconds after the tarp was tied down, it began to pour.

There was thunder and lightning, and the wind blasted the boat. The swell was so calm that the wind seemed to be pushing the boat along the surface. Ron thought it was probably going to push them back to the spot they had started from in the morning. He wondered if that could happen day after day. They could start out and make progress, and then a late-afternoon storm could push them back to the starting point. It would still be worth it, he thought; they might as well be doing something to keep hope and get through the day. If they didn't eventually find land, they could at least stay alive for as long as possible, and maybe a search plane or something might spot them. He thought about satellite cameras and if maybe one of those could spot them going back and forth and work out what was happening. He waddled over to the full water container with his cup and filled it and then tied the tarp down. He drank his cup of water, found a spot to lie down, and tried to sleep.

There was nothing more to do, so he knew he should sleep, but he was wide awake. He thought maybe he had too much sleep the night before, and then the one-hour naps in the afternoon, and perhaps that was why he couldn't sleep at the moment. He thought about all the exercise he had done during the day. He had stayed in the water, kicking along for hours, until his skin had become so water-logged, white, and wrinkled that he'd thought it might peel, and he'd stopped. Then, after his nap, they had rowed at a fast pace and for a long spell. He thought now they probably were moving just as fast in the opposite direction, and he laughed a little. Maybe if they didn't try so hard, the current would take them to Africa in a shorter time than it was taking them to find Australia.

Then he thought Africa wouldn't be good either. Earthquakes, floods, and tsunamis were happening there, too, and it would be no time to be an outsider. There would be enough stress being new to Canberra. He wondered if Canberra's folk would be welcoming. If there were hundreds of thousands trying to move in, maybe they wouldn't be. Then he thought maybe they needed as many people as they could find. It could be so many had died and with so much destruction they might need all the people who could possibly reach there. Maybe they still needed help with disposing of all the debris the tsunami would have left behind. There would be bodies and dead animals, too. It would be a job for volunteers and for the emergency services. The police and firefighter numbers would be decimated and stretched to the limit. He knew his son would be happy to see him, so he would be alright once he arrived, and he felt certain he would survive. He could stay in touch with the others and maybe help them settle there, too, if it was necessary. As his mind continued to ramble, he fell asleep.

When he woke, he guessed it was morning. It was still dark, and the rain was teeming, but it felt like he had been asleep for a long time. He thought it might be very early, a few hours before sunrise. He needed to pee but didn't want to get soaked, so he lay there thinking. He considered it would be easier to get almost entirely undressed and then squeeze out with as little removal of the tarp as necessary. He could have a pee and a shower at the same time. Then he could use his dry shirt he had taken off and left under the tarp to dry and bail out the water that would be on the floor of the boat.

When he moved his leg, his back made a cracking noise. He had forgotten how much his back hurt from the time he'd woken until that moment. He thought it could have been worse than usual this time, and then he recalled how many times he had thought that before. He thought it didn't always make that cracking noise, so that was a change from the recent days, and he would just have to wait and see what happened when he rolled over before he could declare to himself the pain was worse than usual. The pain was over his entire back, and the back half of his right leg and he declared the pain was different but not worse.

When he faced the bottom of the boat, he paused for a few minutes. Whether it was the pain or blood pressure, he didn't know, but he was feeling weak and giddy. He waited for the feeling to pass and untied the tarp. He readied himself and removed the tarp and stood up. He managed to stand long enough to get through the tarp and be almost standing straight before the pain or calcified vertebrae stopped him from straightening any further. He was standing straight enough that he could hang on to the tarp and hold it around himself and to the boat with one hand. That helped him keep his balance while he slowly straightened his back and used his other hand to take aim and make sure he didn't pee in the boat.

The rainwater soaked him, so he dropped his undies, too. He managed to hold the tarp out with almost no rain getting into the boat. He straightened his back and stretched every way he could without letting go of the boat. It was a little choppier than the previous day but nothing he couldn't manage. He opened his mouth and caught the rain and used a hand like a funnel. It worked better than he thought it would, and he drank a liter or so in a few minutes. The rainwater felt great. The saltwater had felt good when he'd been swimming, but it had made him feel kind of crusty when he'd dried out. The fresh water washed that right off.

He rubbed his body all over, trying to remove any dirt or dead skin. It didn't seem necessary with the amount of rain, but he thought it good exercise. He had a sit and leaned right out of the boat and hung on to the tarp while he dipped his bottom into the sea each time the boat dipped. Then he pushed himself up and let the rainwater give him another shower. He climbed back into the boat and found the sponge to soak up the water on the floor of the boat and bailed it out. Then he wrung his clothes a few times before closing the tarp and tying it down, and he crawled over to his jeans and put them on.

He didn't know if anyone else was awake, but it was too dark for anyone to see him if they were. He sat on the seat and concentrated on keeping his back straight. He thought sitting down was usually the worst thing to do, but it would be better than getting into the same position he had been in all night. He couldn't go for a walk, though he could go for a swim when conditions allowed, but that could be hours or days away. He thought if he stayed in one position for too long, he might be forever frozen in that position. When his back started to stiffen as he sat, he moved and flattened it on the seat again. The seat wasn't wide enough for the width of his back. He had slept there all night, but now, where the edges of the seat met his back, the pain was too much of a nuisance to ignore. He spent ten or twenty minutes on his back, then sitting, then sitting and doing exercises. Then he tried sitting with one leg stretched out for ten or twenty minutes, and then the other leg. Nothing really worked, but he thought the exercise had to be a benefit.

Bob and Mary had woken and were at the back of the boat.

Ron said, "Hold on, and I'll move. You guys should have a shower as long as you're out there."

They laughed, and Bob said, "What are you trying to say?"

Ron laughed. "Not that. It's because it feels good; I had one." He motioned to demonstrate and said, "I held the tarp around so I could stand up and not let any water in."

"You didn't let any water in?" asked Mary.

"At times, but it worked pretty well, and the sponge worked well to mop up the floor."

Ron crawled over to the next seat to where Mark and Amy were waking. "Bob and Mary are having a shower. You guys should go next."

They laughed, and Ron added, "Not because you stink, but because the rain feels good. Remember, we said make hay when the sun shines and have a shower when it rains?"

They laughed, and some of the others chuckled. The rain had been teeming, but it slowed to a light drizzle. Everyone had a shower, but it took the late sleepers a lot longer, and the lighter rain would have been less enjoyable.

Ron said, "As long as it doesn't rain any harder than this, we could lift the tarp and sing in the rain as we row."

The lifted the tarp on either end but left the middle of the boat covered. Ron and Steve took the oars, and the others sat in the open. When they were under the tarp, they couldn't sit up completely straight, so when they had the chance, they sat tall and stood from time to time. After a while, the water on the floor was becoming a nuisance. They used their cups to bail it out before it became deep enough to soak the things they were trying to keep dry. The food was in waterproof containers, and things like the barbeque were ok in the water. Their bedding and clothes were kept dry under the tarp they had left covering the middle. When it started raining a little harder, they didn't want to risk getting everything soaked, so they bailed out the bottom and stretched out the tarp again.

They were all under the tarp and had just tied it down as the rain poured harder and harder. They figured they would spend another day under the tarp. It poured for almost an hour and slowed again. They decided against removing the tarp again. It wasn't worth all the effort if it would just start pouring again. It had been an effort to get the water out of the boat to make it nice and dry, and they didn't want everything soaked. They spent the rest of the day and that night under the tarp, only opening it as much as they needed when someone had to pee, and they only did that when the rain would slow. It would pour, then slow down, and then pour again, but it didn't stop raining all day. They were hungry, but the only food they had left were the beans and a little rice. It seemed it would have been impossible to start a fire, so they didn't eat.

Ron said, "If it keeps raining and we can't start a fire, we could just soak the beans overnight and eat tomorrow. On the bright side, we have all the water we can drink."

All day, they had been drinking plenty of water while it was raining. All anyone had to do when they wanted a drink was push the cup out between the tarp and side of the boat and fill it. When it was pouring, it only took a few seconds to fill a cup. Steve suggested they try to fill up on water to stop the hunger, but it didn't really help much. They all made little comments when they filled a cup, like they were having a liquid lunch or wouldn't it be great if they could turn water into wine. They laughed and joked in spite of their hunger and distress and eventually fell asleep.

The next morning, Ron woke up to the sound of light rain on the tarp. The boat was reasonably still, and it didn't sound like any wind was blowing. He thought it sounded like a better day than the one before. When he started to move his legs, he found his whole body was stiff, and the pain in his back was even worse than the day before. He thought that wasn't going to make it a better day, but at least, if the conditions held, he could get in the water and take the weight off his spine. It was slow going, but he managed to turn over, crouch, and untie the tarp before squeezing out to relieve himself. The rain was a sprinkle of a few drops here and there. He stood and stretched for several minutes. The rain felt good on his face and was too light to soak him. He stood as straight as he could and tried to relax.

Steve crawled to the back of the boat and untied a little more of the tarp and joined him. They said good morning, and Steve relieved himself. Then several minutes passed in silence.

Ron said finally, "What do you reckon? Should we have another day of boating on the open sea while the rest of the world struggles with their survival?"

"That's one way of putting it," replied Steve. "I'd rather find land and see what the rest of them are doing."

"Me, too, I'm wondering what's going on in Canberra. They probably think we're among the dead. I suppose we could be spotted by a drone or satellite, but that would be doubtful." After a long pause, Ron added, "It's possible we're on the news and they're looking for us. That would be news worth watching. Too bad we don't have a radio or mobile we could turn the news on."

"The news is probably all about death and destruction. They're probably still recovering and trying to count bodies. Then they will have to figure out who is missing. It's not the sort of news that would help us any, and there has probably been some other catastrophe by now."

"Yeah, we're probably better off out here, and there wouldn't be anything in the news we need to know anyway. Maybe a tsunami warning, but there's nothing we could do to prepare for that, so we're better off not knowing."

"If it was a health warning saying the rain is contaminated with nuclear fallout and don't drink the water, it could be useful."

Ron laughed a little, and Steve nodded and said, "It would be too late if we heard it now and just worry us."

"Yeah, ignorance is bliss. Let's just enjoy our boating while we can."

They smiled and nodded as they looked out to sea in silence. The rain eventually stopped, but the clouds remained. The others woke, and they peeled back the tarp at the back and front of the boat. They left it across the middle in case it started pouring again and they needed to roll it out again in a hurry.

"How about we have beans for breakfast while it's not raining?" said Mary.

They were all hungry after not eating the day before, and they agreed they wanted to eat. Ron agreed they should eat, but he suggested someone could row and someone could swim and push while somebody cooked. Ron and Mark tied a rope around their bodies and jumped in the water. Jin and Julie took the oars, sharing the seat with their children, while Bob prepared the fire and Mary prepared the beans. After a few minutes, the oars were squeaking in rhythm, and Ron and Mark were hanging on to the back of the boat and kicking the water. Ron noticed the weight off his back was giving him instant pain relief, and he smiled wide. Swimming was a good thing because he could get relief and contribute at the same time and the rest of them would never know about the pain he was in.

Mark turned to Ron and said, "Do you get the feeling we move ahead and then, at night, we get pushed back and we're just going back and forth to the same spot and aren't getting any closer?"

Ron laughed and said, "Yeah, I've had that same thought, but I think we may as well keep trying. We need something to do while we're out here."

They kept the boat moving at a steady pace for three or four hours before Mary called out that the beans were cooked. The rowing stopped, and Mark and Ron climbed back into the boat. They took turns to walk or crawl over to the pot of beans, and Mary plopped a cup and a half of cooked beans into their bowls. While everyone was eating, they were all quiet. Ron could hardly believe how good plain beans and salt tasted.

"Man, this is good," he said.

"There's only enough for one more meal," said Mary.

"Let's hope we see land soon."

"Maybe we should have tried to catch fish when we saw them," said Steve.

"I have some fishing gear," said Ron. "A hand line, sinkers, a few lures, and some hooks."

"Let's get it out and try our luck," said Bob.

"As soon as I'm done eating here."

After they finished eating and had cleaned up, Ron found his tackle box. He took out two hand lines and rigged one with a heavy sinker and the other with a lighter sinker and a lure at the end. He suggested they trawl when they were moving along, and they could have two people rowing, two swimming, and two fishing and take it in turns in one-hour intervals. It was a good plan, and everyone else liked it, too. Ron and Bob fished first while Deb and Mary swam and Mark and Steve rowed.

The sea gradually became rougher, the sun was out, and there was a steady breeze that made the heat from the sun a joy. They switched places every hour or so but didn't get any nibbles, much less a fish. Shortly before sunset, the sky ahead changed suddenly, and they could see another storm on the way.

Ron said, "What is this, a storm everyday at sundown?"

They rolled the tarp out and began to tie it down. Ron was at the back of the boat and didn't tie down the tarp in the back corner, leaving just enough room for him to stand so he could remain upright to keep the weight off his spine for as long as he could until the rain arrived. As the storm approached, the wind gained speed, and the water became choppier. There was a huge clap of thunder, and Ron ducked down and closed the tarp and tied it down tight. He informed the rest of them under the tarp that the storm looked pretty bad and they should just try to get some sleep and hope for the best. Then the rain hit, and there was more thunder. The boat was rocking up and down. Ron made himself as comfortable as he could and tried to forget about their predicament and get to sleep. It seemed to take several hours, but eventually, he fell asleep.

He woke and thought it must be morning. The boat was still rocking, and it was still raining. He lay there and thought that at least it didn't sound as bad as it had the night before. Then he heard thunder nearby and thought maybe it was worse than the night before. He looked around under the tarp in the dark, and he couldn't see much, but he sensed that everyone, or most everyone, was awake. They would have been hard-pressed to sleep much longer. The rain was loud enough he would have had to shout to be heard, so he decided not to say anything and lay there with his eyes open for a few more hours. It continued to rain, but he did hear it let up. He turned over and crouched as he untied the tarp and then pushed himself to stand up. He noticed his back was stiff as usual but not as bad as the days before. He felt good about that as he relieved himself.

When he was done, he tied the tarp back down, and then he crawled over to grab the water container. He carried it to the back of the boat, untied the tarp again, and stood up as straight as he could. He grabbed the tarp to hold it up so the rain could funnel into the container. He stood there and stretched for a few minutes before closing the tarp again and carrying the full container back to its place in the middle of the boat. He found his cup and crawled back to the back of the boat. He untied the tarp again and stood in what was becoming his spot. The rain was too slow to merely hold the cup out and have it fill in short time, so he used the tarp to funnel it into his cup and drank it. Then he bailed out what water was on the floor and stood up to fill another cup. He thought he might as well do that until the conditions changed. He stood there for a few hours, hanging on as the boat rocked and almost caused him to fall from time to time.

The others took turns crawling to the back of the boat where he was. While they were relieving themselves, Ron made conversation. He asked them how they slept and tried to joke. He wanted to give them all the impression that things weren't too bad and everything would be alright soon enough. He was having doubts himself, but he thought worry would just make matters worse. He hoped it would stop raining, but instead, it rained harder, and he had to bail out the floor one more time before tying down the tarp and finding a place underneath it to sit still and wait. It poured hard for around an hour. Then it let up for ten or twenty minutes and then started pouring again. That happened all day. When it finally stopped raining, Ron opened up the tarp to see the sun was going down. The boat was rocking less, and the wind had died down.

"How about we leave this much of the tarp up while it's not raining?" he said. "I'll stay here and watch, and if it starts raining, I'll close it up."

They all agreed and again took turns crawling to the back of the boat to relieve themselves. Ron made the spot at the back of the boat his own. He made conversation and jokes with each of them without looking at them as they struggled to hang over the side or stand. When someone would finish and head back to their spot, he would say, "Thank you, come again."

Steve, Deb, and Sue untied the tarp at the front and helped Charlie crawl over and stand up. It looked like he was in a fair bit of pain. Ron thought he looked better the day before but then figured he would be stiff from just waking up. Ron thought Charlie probably had the same kind of pain and stiffness he did, only more concentrated in his ribs. Ron had broken ribs before and knew the pain, and he could see it in Charlie.

When Ron started exercising and stretching, Jin, Julie, and the children came to the back of the boat.

Jin said, "Can we use your hand lines?"

"Good idea" replied Ron. "Help yourself."

They picked up the hand lines and joined him at the back of the boat. Jesse and Eddie took them and let the line out. Jin, with a hand on each line as it passed through his fingers, told them to stop as soon as he could tell the sinkers had found bottom. There was three hundred meters of twenty-five-pound line, and almost half of the spool was out.

"It's at least a hundred meters deep," he said, "maybe one hundred and fifty. We could catch anything down there."

He showed the kids how to lift the lure and then let it go again. They appeared to enjoy it and quietly sat there and worked the lures, rolling up a little line, letting it out, holding it still, and then jigging it, but there were no bites.

Darkness came, and the air was still, and the boat rocked gently. Ron had been standing around for hours, only moving about enough to keep limber but not so much to rock the boat or exert himself. The kids had given up fishing without getting a bite. They all found a spot and tried to sleep. Ron slept width-ways at the very back of the boat with the tarp open. He thought if the rain started, it would hit him in the face and wake him. He could get up and tie down the tarp before too much came in. He used a piece of Styrofoam and his shirt for a pillow and was tired enough to fall asleep.

The next morning, he opened his eyes and saw the sky. There was little or no wind, and the boat was barely moving. He lifted his head and looked over the side of the boat. The sun hadn't reached the horizon, but it was providing light. It looked like it would be a good day, or at least a better day. He rolled over to his stomach and used his arms to push himself up and stand. His back was stiff again. He assessed it as worse than the previous day but better than it had been a few days before that. He thought he would volunteer to swim first and spend as much of the day in the water as he could. Apart from being waterlogged, there was no downside to being in the water. He thought again how it took the weight off his back, gave him exercise to strengthen his back and body in general, and was a contribution, too. They could have more room in the boat, and the boat would keep moving. It was a good thing in many ways.

He relieved himself, washed his hands and face in the sea, and started his stretching exercises. When the top of the sun showed itself over the horizon, he decided to stop and watch it rise.

The others started to wake as the sun was rising. First Bob and then Steve and Mark made their way to the back of the boat and relieved themselves. They said their good mornings and nodded at one another.

Ron smiled and said, "I was just waiting for the sun to rise before I think about getting in the water."

"It will be good to see the sun today," said Bob.

Jin, Julie, and the children woke next, and soon after that, all the girls and Charlie were awake. Ron suggested they untie the tarp and fold it up into a bed again so they could take turns stretching out, and he volunteered to be the first swimmer in the water. They folded the tarp and placed it in the middle-to-front-half of the boat. Sue and Deb helped Charlie lie down on it. It looked like Charlie was struggling with his pain again. Sue helped him hold his head up to have a drink of water.

Ron was looking at the sunrise while Bob stood next to him looking in the other direction. Bob called out for him to have a look at something, and Ron turned around. He could only see birds over the water. It was too far to see what they were interested in, but there were a lot of them.

"What is it?" asked Bob.

"We're going that way anyway," replied Ron. "Let's see what it is."

He found his cup and drank two cups of water. Then he tied the rope around himself and jumped in. Mark joined him, and Bob and Steve started rowing. They made steady pace for aboutan hour before Mark called for a break. They all looked ahead and at the flock of birds hovering over the water in the distance.

"It doesn't look any closer," said Bob.

"I know," said Ron, "but it is, so let's keep going. Who wants to join me for a swim?"

Bob and Steve were sweating profusely. It was still early, but it was humid, and the sun was beating down. Bob and Steve both volunteered for the swim.

"Fair enough," said Ron. "I'll get out, and you guys can have a swim."

Ron and Mark climbed into the boat, and Bob and Steve jumped into the water, and Jin and Julie took the oars. Sue and Deb had been lying down either side of Charlie.

Sue said, "You guys come and lie down and get some rest."

Ron didn't feel he needed rest, but he thought it was a good idea to lie down and stretch out while he had a chance, so he agreed, and he and Mark lay down on either side of Charlie. The sun was bright, so Ron took off his shirt to cover his face and eyes. He was wide awake when he closed his eyes, but he fell asleep within minutes.

A few hours later, Ron woke to see Sue and Deb at the oars and everyone else crowded from the middle to the back of the boat. He layon his back for a moment and thought about how good it was feeling after the swim and a stretch. He turned over and stood up slowly. Some of the others were pointing ahead and encouraging him to have a look. He turned and saw they were getting near the birds. The birds were working the water around the edge of a huge mass of debris or rubbish. He wondered if it could be the same pile of rubbish they had seen the other day. If so, that might mean sharks were about. It could also mean they were back to where they were before, which could mean they were nearing land. Then he thought it could be a different mass of rubbish floating around the sea, or if it was the same one, maybe it had moved further out to sea. Either way, it would be worth having a look at.

Mark woke up, too, and Bob and Steve made their way to the front of the boat. Sue and Deb stopped rowing, and everyone was looking ahead and around the boat for sharks. Ron and Mark made their way to the oars and started rowing. Half an hour later, they could see what appeared to be an island of trash with birds working the edges. There were a few sharks that swam past the boat. Ron and Mark stopped rowing.

Ron said the sharks weren't as active as the last time and asked if they wanted to get closer or not. Bob and Steve said yes, but Mary, Sue, Deb, and Amy shook their heads no, and there were some other moans of doubt. They decided to take it slow, and if it looked like trouble, they would go around. Ron noticed what he thought might be salmon or tailor breaking the water.

"We might be able to catch a fish or two now," he said.

Little Eddie perked up, smiled, and said, "I'll fish."

Ron nodded approval, and Jin helped him dig out the hand lines. They rigged one with a typical steel lure with three treble hooks. Ron continued rowing while Jin started giving Eddie advice about setting a hook when a fish bites it. Then Jin rigged another hand line for himself.

They approached the floating rubbish pile slowly, stopping every few minutes to have a good look around the boat. They saw a few sharks, but the sharks didn't appear to be interested in the boat. Jin caught something on his line, which surprised everyone. He started to pull the line in hand over hand, and it appeared it was something heavy and putting up a fight. Little Eddie watched him with intent and his mouth wide open. Then Eddie got a bite, too. It startled him, and the line started rolling out as he dropped the spool, which hit the floor and started spinning as he tried to pick it up again. Julie came to his aid and picked it up.

Steve said, "Bring them straight up, or the vibrations might attract sharks."

Julie reacted with a look of fear as she started pulling the line in hand over hand. Jin got his fish into the boat, and it was a salmon around a foot and a half long. It started flopping around the bottom of the boat before he got his hand over it and picked it up with his fingers by the gills. Ron took out his pocket knife and tossed it to Jin and told him to cut off the head so it would stop flapping. Jin cut the bottom of the throat and removed the gills with the head still attached.

Then Julie pulled in the fish Eddie had hooked. It was a salmon just slightly smaller than the first one. It was thrashing about, too, so Jin cut it and removed its gills like the first one. It was all very exciting when the fish was thrashing. Eddie was clapping his hands, and Jessie was right there with him with her mouth agape. When the fish stopped moving, so did Eddie and Jessie. A shark approached the boat, and there was a quiet groan of apprehension and shushes.

Ron whispered, "Nothing much going on around here, Mr. Shark," and the kids giggled.

They were all quiet and as still as possible. The shark went under the boat, and they lost sight of it. Several silent minutes passed before Jin asked if they should start fishing again.Ron said the shark would have probably scared the salmon away, so maybe they should get closer to the pile of rubbish and see what it was that had so many birds riled up.

Then he and Mark rowed at a steady pace while being very careful with the oars so as not to splash the water and attract the sharks. They lowered the oars and gently took a long stride, then gently lifted the oars out of the water, and then they slowed down to a stop before gently placing the oars in the water again to repeat. There were more sharks as they neared the rubbish island, but the sharks showed no interest in them, so they kept moving. They reached the edge of the floating mass. Most of the rubbish looked like plastic bottles in the thousands, or maybe millions. There were advertising signs, small boats, and surfboards among the floating debris. When they were at the edge of the pile, they couldn't see where it ended on either side.

Ron said, "This is huge. It must be all the rubbish caught up in a current after the tsunami."

Then he noticed something huge in the rubbish that looked like one solid piece. He looked closer and pointed for the others to have a look. He guessed it was the ass end of a container ship sticking out of the water. Then they all noticed similar lumps in the rubbish and some that were longer. Wherever there was an elevated bump on the rubbish pile, it appeared to be a partially submerged ship.

Steve pointed and said, "Look over there; that looks like a passenger ship."

A few hundred meters into the pile, they could make out windows and decks that looked to be the top floor of a cruise ship.

"This pile of rubbish is a graveyard for ships," said Bob.

"Probably people too," replied Ron.

They entered at the edge of the pile, and the boat started to separate the floating plastic bottles as they pushed in for five or ten meters. They feared if they went too far in, they might not be able to get out. They turned the boat around to get back to clear water and began circumnavigating around the island of rubbish. Ron spotted what looked like a sheet metal door with a timber frame close to the edge.

He said, "Let's go pick up that door; we can use the timber."

They pushed the boat through the plastic bottles and picked up the door. When they were pulling the door into the boat, they spotted a Styrofoam body board among the rubbish and picked it up, too.

Ron said, "We can start a fire on the sheet metal and keep it afloat with this. We won't need to dig out the barby. The wood is pretty well waterlogged, but it will dry out soon enough. There's more than twice as much as we have."

They decided it was a good idea and time to stop and cook the fish they were all keen to try. Ron picked up the last of the timber they had saved and put the waterlogged timber they had just found in its place. He tied a rope around the body board and the other end to the boat. Then he placed the sheet metal on the board and started the fire on top of it. He plucked an empty plastic bottle out of the rubbish and asked Eddie to scale and gut the fish and put the guts in the bottle so they could use them for bait.

Mary filled the pot with fresh water, and Ron put it on the fire. He put the lid on and checked under it every few minutes. When the water began to boil, he called for Mary to drop the fish in. They had to cut each fish into thirds to fit them in the pot of boiling water. Mary dropped the fish in the pot, and Ron put the lid on top. He waited for what he thought was three minutes and used his shirt for a pot holder and was going to bring the pot into the boat. The shirt didn't work too well, and he burned his fingers, so he put the pot down quickly. He grabbed another empty plastic bottle, filled it with seawater, and poured it onto the side of the pot to cool it down. Then he soaked his shirt before using it to grab the pot and bring it into the boat. He used his wet shirt to hang on to the lid as he poured the water out over the side of the boat while holding the fish in the pot with the lid. When he was done, it looked like a nice pile of meat in the pot.

He said, "Bring your cups, and be careful not to choke on any bones."

He filled each cup, and there was almost half left over. When everyone finished eating their first cup, they came back for another. Ron was dispensing the food, and he guessed the second helping would be about three-quarters of a cup, which it was, as if he had measured for perfection. He said the portions were just about perfect and the chefs should get a compliment. There was a grunt in the background, and the general consensus was it tasted kind of plain but was a great feed. It was late afternoon when they finished eating.

Ron said, "What do we all reckon we should do?" There was no answer, so he asked, "Do we want to have a look around here or keep moving to where we hope land is?"

"Someone might spot us here," said Bob. "They might be looking for the ships."

They discussed the pros and cons. Maybe they were near land, and maybe someone would be looking for the ships or to recover bodies. Maybe they wouldn't be getting any nearer to land and might be able to catch more fish if they stayed put. They decided to have a look around for the rest of the day. They would fish until dark and see if there was anything else in the rubbish pile they could use for firewood.

Ron removed the lures from the hand lines and replaced them with hooks. He put a hook through some of the fish guts Eddie had saved in the plastic bottle. Then he dropped it over the side. He waited for it to touch bottom, and then he lifted it around two feet and held tight. He figured he was fishing eighty to one hundred meters deep.

"Who wants to join me?" he asked.

Little Eddie stepped up without hesitation and with a big smile on his face. Ron baited his hook, and little Ed dropped it in the water and let the line roll off the spool. As Ed's line was on its way down, Ron got a bite and set the hook. He started pulling it up hand over hand, and a few minutes later, he pulled in a little trevalla around a foot long. He said it wasn't huge but would be good eating and held the fish to remove the hook. Eddie had let his line down as he was lifting it off the bottom, and he got a bite and set the hook. There were cheers and a feeling of excitement.

Ron asked Julie for the bucket she had used before when they'd been washing. She brought it over, and he threw the fish into it. He baited his hook and dropped it in again. Eddie brought up his fish, which was the same only slightly bigger. Jin helped him get the fish off the hook and put it with the other one. The two fish were flapping, so Jin cut them under the head and removed the gills. Ron asked him to scale and gut them, too, so theycould use the guts for bait while they were fresh. Then he got another bite and set the hook. He was pulling it in hand over hand when Mark came crawling to the back of the boat.

"That looks like fun," said Mark

"You can go next," Ron replied.

As Ron was bringing his fish into the boat, Eddie was dropping his hook over the side again. When Ron unhooked his fish, he gave Mark the hand line and called for anyone else who wanted to fish to take turns at the back of the boat.

He crawled to the front of the boat and found a spot to flatten his back on the floor. The sun set, and they were still fishing. The bucket was nearly full of scaled, gutted fish with the gills removed and the heads attached. The plastic bottles with fish guts were full, and they had been throwing some over the side for a while. They decided it was too dark to keep fishing and laughed when Ed said not many more fish would fit in the bucket if they caught any more anyway. Ron made his way back to his spot at the back of the boat, cleaned it, and soaked up what water was around with his shirt. They covered the boat with the tarp, and everyone found a space to get comfortable and try to sleep. Ron thought they were all in good spirits, and that was incredible when he considered how long they had been lost at sea. He closed his eyes and was soon sleeping.

Chapter Five
A Lot of Rubbish

The next morning, Ron woke and opened his eyes. It was still dark, and he remained still while he recalled what had happened the day and night before. The boat was rocking more than the day before, and he could hear the wind. They had plenty of fish now, but that didn't mean they would have conditions to cook. He thought about whether they should stay around the pile of rubbish, where they could probably catch fish and maybe survive long enough for a search plane or passing ship to happen by, or if they should keep moving. He thought they couldn't hang around Rubbish Island indefinitely, so of course, they should keep moving and hope to find land. They had food that should last three or four days, and it shouldn't take that long to find land, he hoped. They could get around the edge of the rubbish and look for more firewood on their way out.

He rolled over and crouched as he untied the tarp. He stood up slowly and realized his back was worse again. The boat was rocking more, and then itlurched, and he had to hang on to keep from falling. He was hanging on tight while he relieved himself. It was still dark, and he couldn't see a thing. He removed more of the tarp from the back of the boat and stood there, trying to keep his back straight, and did some stretching.

As the sun showed, the others started to wake and move around. They had drifted away from the pile of rubbish, so Ron and Mark took the oars and started rowing. About an hour later, they neared the edge of the pile at about the same spot they'd approached the day before and stopped rowing.

Ron said, "You guys know I have a mask and snorkel."

"Do you want to go for a swim?" asked Bob. "There are a lot of sharks around here."

"I might just take a dip, have a look, and then get back into the boat."

Bob took the oar from Ron and helped Mark keep the boat near the pile while Ron dug out his mask and snorkel. He tied a rope around his ankle and the other end to the boat before he dropped over the side. He dived under a mass of plastic bottles, bits of Styrofoam, and other rubbish. The visibility was great, and what he could see was the deck of a cruise ship with two container ships on top of it. The containers had all collected at one end of the container ships and were askew in huge piles.

He also saw a trawler, a yacht, cars, bits of timber, and sheet metal. He was about twenty feet below when his rope became taut. He looked up and could see underneath the floating rubbish. He saw a large white pointer swim towards him, and he grabbed the rope hand over hand and was at the side of the boat in an instant. He called out for help to get in, and Bob and Mark grabbed his arms and pulled him in the boat. He had only been in the water for a matter of seconds.

"That didn't take long," said Bob.

"I saw all I needed to," replied Ron. "A huge white pointer."

At that moment, the shark came to the surface, and Mark yelled out and pointed as everyone in the boat looked and held their breath. They all froze and remained perfectly still while the shark swam past, and then it circled and returned a few times.

"It really is like a ship's graveyard," said Ron. "There's a cruise ship with two container ships on top of it, and another cruise ship deeper in."

When the shark returned, he went quiet and when the shark swam away, he continued.

"There were other boats, too, and an expensive-looking yacht and a trawler. If we were better prepared, or there weren't sharks about, we might be able to salvage all kinds of things."

They discussed what they might find if they could open the containers or explore the cruise ship. When the shark returned, they went quiet and remained still. Once it swam away, they resumed talking. Of all the things they thought they might find, none of them would be particularly useful while they were in the boat. Maybe if they found gold or diamonds, they could fit enough into the boat to make it worth their while, but that would be an unlikely find.

They decided they would stay close to the pile and go around it. If they found some timber near the edge, they could collect it. The plan was for them all to keep their eyes open, and if anyone saw anything useful, they should speak up. They spent the day getting to the edge and venturing a few meters here and there into the pile while moving around looking for something worth having. They managed to find a lot of timber and stopped collecting it when they decided it was more than enough. There was only so much they could fit in the boat and still have room for everything else, including themselves. What they were unable to tie down under the hockey straps, they left untied. They also found a few useful pieces of Styrofoam, a spool of nylon rope, and a surfboard. They towed the surfboard with some of the nylon rope; they intended to use it cook on. There was little room on the boat, and if they could make the fire on top of the surfboard tied next to the boat, it would be a lot easier.

They would have no trouble finding a piece of sheet metal. The trick would be finding a way to make it stay on the surfboard. They decided to stop and try it while the conditions were calm. It was a few hours before dark, and they were all hungry. They held the surfboard close to the boat and put a square-meter piece of sheet metal on top of it. Ron dug out the timber from the day before and started to splinter the edges. Steve took the splinters to start the fire while Mark held on to the sheet metal to make sure it stayed on top of the surfboard.

Ron said to Mark, "That's going to get hot."

"I can use a wet shirt or something," replied Mark.

"Ok, girls," said Ron, "we're having a wet t-shirt competition, and whoever has the wettest t-shirt can take it off and give it to Mark so he doesn't burn his hands."

There was quiet laughter, and Bob said, "Here you go. You can have mine." He stood up and took off his shirt before doing a little shimmy dance that brought laughter.

They decided to cook all the fish while they had comfortable conditions. Not all of the fish would fit into the pot, so they cooked as much as would, and then they emptied the pot and started eating while they cooked the rest. It was dark before they finished. They had all they wanted to eat and enough cooked fish leftover to eat cold for another meal the next day. The conditions were still calm, and because it wasn't raining, they decided to sleep under the stars. There was still timber they couldn't fit under the hockey straps, so they took most of it and put it on the fire. It entertained them and kept them warmer for hours as they stared silently at the flames and then the hot coals.

They fell asleep for four or five hours before the wind picked up and the swell started to rock the boat. The boat suddenly rocking woke them all. It wasn't raining, but the weather had certainly turned. They relieved themselves before stretching out the tarp and covering the boat. They found their spots under the tarp and made themselves as comfortable as they could. The boat was up and down for half an hour, making it impossible for anyone to sleep, and then they heard the rain. They repeated the saying that was becoming their mantra: when there's nothing else they could do, they should get some sleep and hope they would still be there in the morning. The boat was rocking at a fast rate, and the sound of the rain was so loud they had to shout when they said anything.

Ron managed to sleep a few more hours. Then he woke up and listened to the rain. He wondered if anyone else was awake. There wasn't much else they could do other than lie there, so he tried to sleep. He would fall asleep for a few minutes at a time, and then he would be awake for another twenty minutes. He had his back flat on the floor, but he didn't have enough room to straighten his legs. The pain in his back became progressively worse, and his muscles and joints felt stiff. He had a different kind of pain under his legs, and when he moved them, one cramped up. He couldn't move it, and it hurt like crazy, so he remained as still as he could and tried to relax. He wondered if anyone else was having the same problem he was and how they were putting up with it. He thought they must be having similar pain. How could anyone not have pain when they were stuck all bent up in the same position for hours at a time. Then he thought, *What if it goes on all day, or for days?* He would just have to put up with it. There was nothing else he could do but hope it would stop raining.

After a few hours of constant rain, he heard it let up. Several minutes later, it started pouring again. It would pour and then let up several times during the day, but it didn't stop raining once. He needed to pee, and he waited for the next time the rain slowed down. As soon as it did, he made his move. His stiff back made him slower. He could ignore the pain, but he couldn't make himself straighten up any faster. He managed to crouch to untie the tarp, and then he stood up and relieved himself. It was light rain, but in the time it took him to finish, he was soaked. He wrung out his shirt, and then he couldn't find his sponge, so he used the shirt to sop up the water on the floor. He was about to re-tie the tarp when Bob came to the back of the boat.

"Me next," he said.

"Ok," said Ron, "I'll leave my shirt here so you can mop the floor when you're done."

Mary had made her way to the back of the boat, and she said, "Make room for me."

"Ok," Ron replied, "I guess we will all have to go. I'll leave my shirt here so when everyone is done, they can use it to mop the floor."

He crawled to the front of the boat. Charlie and Sue were there, and Charlie had to pee and was in a lot of pain. Ron fetched an empty plastic bottle he had saved and handed it to him.

Ron said, "I don't know why I didn't think to use that in the first place."

"It's ok for you guys," said Sue, "but we would need to open the tarp anyway."

"You could use Julie's bucket we had the fish in, but then we may not want to put more fish into it."

Sue laughed. "I hope we find land before we need to catch more fish."

"Me too."

When everyone had finished, the rain was teeming again. They huddle under the tarp and figured it must be late evening or early night. They were all hungry, and they happily agreed to eat the rest of the fish. It was enough for everyone to have a full cup, and they didn't mind one bit it being cold.

Ron said, "I wouldn't say it tastes good, but it's good enough."

"If I was at a fancy restaurant," said Bob, "I would send it back, but I'll be keeping this."

Ron laughed and said, "Imagine if they served this in a restaurant."

"If they said cold boiled fish was the new fad diet," said Sue, "people would eat it."

"If they were as hungry as we are," said Ron, "they might think it tasted alright, too."

They had to almost shout to speak, but the conversation and jokes started rolling back and forth from one end of the boat to the other. They talked and joked about what they might eat if they were in a restaurant and what other people in restaurants might think about eating plain boiled fish without anything else, or plain beans. The rain kept pouring, and the boat rocked until the conversation tapered off and they started to fall asleep. It was difficult to sleep, but they were so tired they decided they should all try. Ron made his way to his spot at the back of the boat and closed his eyes.

Throughout the night, they slept in patches. Several times, the boat nearly tipped over, and none of them could sleep through that. It had stopped raining at least twice but not for more than ten or twenty minutes. Ron opened his eyes to darkness and the sound of rain on the tarp. He stared into the dark, at the tarp just a few feet above him, and listened to the rain. He reckoned it was slower than the night before. He thought he would just remain still until it slowed down and then get up and pee and start stretching. It was becoming a morning ritual, just like chores used to be at home. He thought about his house and wondered what had happened to it. It was probably underwater if it was still standing. It might be completely underwater, or maybe the roof was above water. Maybe he could go back and live in the attic. He laughed at himself for having the idea and thought it was as dumb as the thirteen of them living on this boat for the rest of their lives. Then he thought that might be possible, but only if their lives were short.

He told himself to not despair, to keep positive. They would get out of this mess and survive. After a few hours, he heard the rain slow down. He thought about waiting to see if it would stop altogether, but then again, it could start raining harder, and he would miss his chance to get out and pee. He compromised and decided he would move slowly. He rolled to his stomach, and when he started to get up, he found out how much pain he had. He thought about what he could do to stop the pain, and the only thing he could think of was to stop moving, and that didn't help. He wanted to scream when he crouched to untie the tarp, but he didn't want to wake anyone, just in case anyone was managing to sleep. He opened the tarp only as much as he needed to squeeze out and stood up. The boat was rocking, and he needed both hands to hang on to the side. He looked out and could see they had again drifted a long way from the rubbish pile overnight.

There was enough light to see the clouds, and he thought the day looked promising. It wouldn't likely be a sunny day, but it might stop raining. The rain wasn't enough to bother him, but the boat rocking the way it was would make it difficult to stand and pee. He ducked back under the tarp and grabbed the empty plastic bottle he had saved for the purpose. He stuck the top half of his body back outside of the tarp. With one hand, he held on to the side of the boat, and with the other, he held the empty bottle, with a finger to hold him in place so he wouldn't need to aim. It took skill and dexterity even though it was a simple and compulsory task, and in his mind, he congratulated himself for not spilling a drop. He emptied the bottle and managed to rinse it with sea water when the boat was dipping.

The rain stopped, and he stood there hanging on to the side of the boat. It was rocking too much for him get any value from stretching, so he just tried to keep his spine straight. The others were awake, and when the rain stopped, they stirred and removed the tarp, leaving just a meter or so to cover the middle of the boat.

"I thought we were going to tip over a few times last night," said Bob.

"Yeah," said Ron, "me, too."

"We should probably grab the oars and keep nosing into the swell before we get swamped," said Steve

Ron pointed. "Rubbish Island isn't too far that way, but I don't know which other way we drifted."

"We should just stay upright and worry about which direction to go when the sun comes out," replied Steve.

They had different opinions about when the sun might come out and which direction they might have drifted in the night and decided they would have to wait for the sun to show itself to know for sure. Steve and Bob took the oars. They set about facing into the waves and trying to keep the boat from smashing the bottom of the waves or tipping over. They soon found a knack for it, and the boat was gently rolling over a three-meter swell. Keeping the boat from rocking any more than it had to became the focus. There was nothing else they could do, so they all watched intently and gave encouraging moans when they managed to get over a large wave with little fuss.

That kept them entertained for the entire day. They took turns with the oars, and with every new pair, there was a learning curve. At first, the boat would drop and hit the water, jarring them as they hung on. Then each pair quickly got the hang of it, and it became a skill that gave them a sense of satisfaction.

All the action had them very tired but in good spirits when the day ended. They were able to see the sun as it was setting, and they enjoyed the moment. They would at last know for certain which direction they were going for a few minutes. Jin and Julie were at the oars, and as well as making sure they faced the waves, they headed in the direction of the setting sun. The swell was too large to see very far, and they had lost sight of the rubbish pile. There was still a lot of cloud as the sun set, so it didn't take long before darkness was on them again. They stretched the tarp out and covered the boat. As soon as they were under the tarp, the boat took a big drop and splashed at the bottom, knocking them off their seats.

Ron said, "That's the difference between drifting and using the oars. It's going to be harder to put up with it tonight after we had it so smooth all day."

Bob laughed and said sarcastically, "Yeah, today was so smooth I can hardly feel the blister on my ass."

They all found their spots and made themselves as comfortable as they could for the night. The next morning, Ron woke up and opened his eyes and thought it was the same way things had been the previous morning. The boat had been rocking all night, and again, he felt lucky they hadn't been tipped over. He listened and thought it better than the previous morning because it wasn't raining. He rolled over to his stomach and thought his back was much worse than the previous day. Then he thought about how often he had considered that same thing, whether it was yesterday, last month, or years ago. He struggled to get up and to do all the things he had done the morning before.

The others were awake, and as soon as it was light enough, they had removed the tarp, and Ron and Bob took the oars. They again found the knack of rolling softly up and down the waves, this time without hesitation. What a pleasant difference it made when the boat wasn't crashing about from time to time.

The day was a little brighter, so they could edge in the direction they considered to be nearest to land. They had lost sight of Rubbish Island. They took turns with the oars in the same manner they had the day before. One other thing that was different was when the two people were rowing, the others tried to sleep. It was easier to sleep during the day while the boat was reasonably steady, without the occasional crashing that happened during the night when they were adrift.

At the end of the day, as the sun set, Bob said to Ron, "The day went fast."

"Yeah, replied Ron, "it's intense when you have the oars in a big swell. It gives you something to do, and the time flies."

They all made sure everyone was ready to get under the tarp for the night and then tied it down. Ron flattened his back on the floor and felt pain shooting up and down his back. It felt like his back was moving. He thought his back pain was a sensation because the muscles under the shoulders and along his lower back moved involuntarily. He figured he could say he was sensational, and he smiled and remained still while his muscle tics entertained him. Eventually, he fell asleep.

The next morning was like a carbon copy of the previous one. The night had started as rough as the previous two, but they had all become used to it. They had been tired and hungry, but they knew they had enough food for another meal as soon as conditions allowed them to cook. They agreed to not talk about food until it was time to eat. As they were moving about and opening their eyes, they were all hungry and hoping to eat soon. The rain poured, and their hopes were dashed as they rushed to stretch the tarp back over the boat. They all did their best to adopt their new notion on life.

"If there's nothing else we can do, at least we can stop worrying and take it easy," said Ron.

They laughed and made more jokes, like, "If you're falling from a plane without a parachute, stop worrying and get some sleep."

Another: "Whether in a tornado or volcano, just close your eyes and see if you're still there when you open them."

And:"When an earthquake starts, it makes it easier to move your bowels."

Ron said, "You might be able to run from a volcano, and if it's an earthquake, get under the table, but if you're in a tornado, you might as well stay where you are, and ideally, you are in a bar getting drunk."

That started jokes about country and western songs and homes going underwater or being torn up in a tornado. They had to relay the jokes from the back of the boat to the front, and when those in front heard the joke, the people at the back of the boat knew it because they heard their laughs. When a joke came from one end at the same time as one from the other end of the boat and they met in the middle, the people in the middle got two new jokes and then passed them on or added one. The laughter rippled back and forth from time to time throughout the day until they were tired enough to sleep.

The next morning, Ron woke, and as he lay there, he thought about what good spirits they were all in and the role he was playing. The others were all partners, and Jin and Julie had Eddie and Jessie. He was the only one all alone when they all paired up, and he was the one, it seemed to him, who was always trying to cheer the rest of them up and tell them everything was going to be ok. He thought he was being helpful to the group and would just keep doing whatever the next thing required, climb a mountain or molehill or whatever it was, until he was in Canberra with his son and granddaughters.

The boat had settled considerably, but it seemed to be moving. It was still dark, but he was wide awake and wanted to have a look at what sounded to be calmer conditions. He went through his ritual with his back and considered it his new routine. The pain was bearable, but his body just wouldn't bend when he wanted it too. It seemed it was taking more time each morning. He squeezed through the gap between the tarp and side of the boat and stood up slowly. He was back to the one vertebra at a time, and it took a minute or two for each one to straighten. It took him at least twenty minutes before he could stretch for the sky and move his neck from side to side. He was giddy and didn't know if it was a reaction to the pain or the sea. He remained standing, hanging on to the side of the boat as he felt pins and needles and then numbness. The boat wasn't rocking much, but it felt like it was moving rather fast. He concentrated on his breathing and tried to take deeper breaths. The time and odd feeling passed, and he decided to relieve himself while the boat wasn't rocking. He took a seat over the side of the boat. It wasn't easy, but he thought at least it would probably be the last sit over the side he would need until after they ate again, and he sure did want to eat again.

When he reached into the seawater to rinse his face and hands, he noticed the boat was moving and rather swiftly. There was just a hint of light, and he thought he could see circles like there might be in a river current. It wasn't raining, and he knew it would be light soon, so he opened up more of the tarp. He was certain they were in some kind of rip or current and were moving faster than they ever could have with the oars. Steve, Mark, and Bob woke, and Charlie and Sue were at the other end and opened up the tarp. Ron gave them all the lowdown: they were in some kind of current, but he didn't know what direction they were going. They all went through their own morning rituals much like Ron had an hour or more before.

It was cloudy and gray, so there was no sunrise, but they could tell where the sun was, and it was in front of them. That meant they were going in the wrong direction to reach Australia.

"We may be closer to New Zealand anyway," said Bob.

"Don't forget, New Zealand probably isn't there anymore," replied Ron.

The boat either picked up speed or it appeared that way when they had enough light to see. It was like they were in some moving current after all, and the water was moving in the one direction for as far as they could see from either side of the boat. The water became increasingly rapid, and white caps popped up from time to time.

Ron said, "I don't know what this is, but it's something."

"I don't like it either," said Bob.

"Do you think we ought to go to sleep," asked Steve.

That cracked everyone up. Charlie hadn't heard it from his end of the boat, and when Sue told him, he laughed and grabbed his ribs, the pain showing on his face. Ron laughed at that, remembering what it's like to laugh when you have broken ribs. It hurts, but it feels like it's doing you good,like laughter really is good medicine, so he thought it was good Charlie was laughing.

He looked to the front of the boat and could see a wall of water that might have been a hundred meters tall. It was far away, and with boat picking up speed and rising in the air, he couldn't estimate how far away it was or if it really was as tall as he'd first thought. They looked ahead at a huge wave, and when they looked back, they were looking down thirty or forty feet of foaming whitewater. There were shouts of horror and fear. They kept rising, but luck was with them because the boat did not get vertical. At the highest point, Ron thought they must have been at least a hundred meters in the air. He looked to the back of the boat, and in the far distance, he could see the rubbish pile, and he pointed it out.

"Look, he said, "there's Rubbish Island. From way up here, it doesn't look that big. Well, at least we know we can get around it and there's another side."

The others looked, but the predicament they were in and their fear had their immediate attention, and nobody else passed comment on the view of Rubbish Island. The descent on the other side of the wave started out a wild ride at fast speed, but it leveled out and left them bobbing before settling to a calm sea. There was fear and excitement and then relief and almost disbelief as they stopped moving.

"How long do we reckon this is going to last?" asked Ron.

They looked around the boat and saw calm water on the horizon in all directions. Behind the boat, they could see the wave in the distance. It looked like the horizon was moving.

"I'll bet that's another tsunami heading for Sydney," said Bob.

"We lucked out again," said Ron. "It's a miracle we weren't swamped. It would have been even better if we could have caught the wave all the way to Sydney, but then we would have needed another miracle."

"There are only so many miracles to go around," said Steve.

They cheered and laughed loudly and agreed they were again lucky to be alive. Everyone in the boat was relieved and happy. They were laughing, shaking hands, and hugging, and they decided it was an omen they were meant to survive and that they would survive. They watched as the moving wave on the horizon moved further and further away.

Half an hour later, the water was still calm, and they decided to eat. It would be the last of their beans and less than they had eaten the previous time. Because there was less and just a small amount of rice left, they threw in the rice, too. There would be enough for each of them to have a cup of beans and a few spoonfuls of rice, and then there would be no more food in the boat. They talked while they cooked what would be their last meal unless they were rescued or found food out in the sea. The water was still, and the wave on the horizon was so far away they only knew where it was when they looked around the other directions. Everywhere else, the horizon was flat, but in the direction of the wave, it had variances. It had turned into a perfect day with mild conditions, and Ron was keen to get in the water and get the weight off his back. He suggested they start moving again and volunteered to be the first in the water.

"I'll take the first leg in the water if anyone wants to join me," he said. "Who wants the oars first?"

They agreed they needed to keep moving. At least they were pretty certain about which direction to go. They were worried about being out of food, so they were really hoping to find land soon. Ron suggested somebody should have a lure in the water while they were moving, and Eddie quickly volunteered for that job. Ron and Steve took the first spell in the water while Bob and Mark took the oars.

The rest of the day went well, and at sunset, it was still calm and wasn't raining. Ron had spent about half of that time in the water. He had also done some rowing and had slept for a continuous hour, which had been difficult to do during the previous few days or nights. He was feeling great and looked around at the others, who all looked to be in good spirits. Even Charlie looked better. At times, it was obvious he still carried some pain, but other times, he was moving around on his own. Charlie had spent an hour or more in the water, and Ron guessed that would have helped him a lot, too.

They had been keeping a steady pace, and Ron concluded it was a good day. They had survived with the help of a minor miracle, had something to eat, had a swim, eased their pains, and made progress towards land, knowing they were going in the right direction. He thought, *What else could be better, except to wake with the boat landing on a beach?* They covered most of the boat with the tarp and left the rest open so they could keep moving until dark. When it became too dark to see, they stopped and covered the rest of the boat and made themselves comfortable for the night.

The next morning, Ron opened his eyes. It was still dark, and he recalled the previous day and miracle and wondered what he would see when he poked his head out of the tarp. He could hardly wait and was hopeful he would see land. He thought surely they had been through enough and this ordeal would end soon. He knew nothing they had gone through would make a difference in regard to finding land, but he could at least hope. He rolled over to his stomach and got up to untie the tarp. Then he slid out and stood there stretching as his eyes tried in vain to see through the dark. His back was better than the morning before, but he still needed the stretches. He also wanted to be the first one in the water again, and he wanted to get in as soon as the sun was up.

The boat was barely moving, and the breeze was cool and mild. It was still pitch black when he finished his stretching. He untied more of the tarp to open up a little more space. He took a seat over the side of the boat, rinsed his hands in the sea, and made his way to the container to pour himself a cup of water. The water container was still over half full, so it wasn't really a worry, and everyone would certainly have a cup as soon as they wanted. Then they would probably have another cup during the day and then another before sleep and maybe even more because there was no food. He thought that would see them out of water in around three days, so maybe they should cut back the second day, and if not the second, then definitely the third day. Hopefully, they would find land before then or it would rain.

The others started to wake, and one by one, they made their way to the back of the boat to do their morning business of relieving themselves. Then they crawled back to their spots to wait for the light. It started to lighten but not so much anyone would see the sun. Ron, Steve, Mark, Bob, and Jin were all at the back of the boat, trying to see the daylight. It was gray and cloudy, but from the lighter sky behind and to the right, they knew where the sun was. Steve pointed in the other direction and injected some enthusiasm as he called for them to get moving. They removed the tarp up to the front seat, where they left it rolled and ready to cover the boat again in a hurry if the need arose. Steve and Mark took the oars while Ron and Bob tied a rope around their upper bodies and dropped over the back of the boat and into the water. They couldn't see much, but they knew they were going in the right direction. It was calm, and they were feeling strong.

They sped up the rhythm to the point Ron and Bob were hanging on to the back of the boat and kicking was slowing them down. They eventually found the sweet spot where they could all contribute and make time without tiring. They went for more than two hours at a steady clip until Bob asked for a break. Ron liked the idea, too, so they climbed into the boat while Jin and Julie dropped into the sea. Mary and Amy wanted the oars, which prompted jokes about women drivers, but they eventually took them and started rowing. It took a little longer for them to find their rhythm, but soon enough, they were making good time, too. The clouds were high and gray, so it didn't look like rain anytime soon. They were all in good enough spirits and satisfied with the day so far. Making fast time in the right direction with the oars was easier and more satisfying than using the oars to keep the boat from being swamped in a huge swell.

"It's going well so far today." Ron looked ahead and suddenly saw something on the horizon. "So, here something comes."

They couldn't tell what it was, but they knew it didn't look like land.

"We're looking towards land, so how could it be a tsunami?" said Ron.

"Should we close the tarp up?" asked Bob.

Ron shrugged. "Should we leave it down so we can see in case we need to steer over waves or whatever this is?"

They looked at each other, and Bob said, "We don't have long to decide."

"Anyone have any ideas what we should do next?" asked Ron.

Jin and Julie were in the water, hanging on to the back of the boat, and Jin said, "Help us get back in the boat."

"Ok," said Ron, "but what after that?"

They all went silent while they watched the changing water charging towards them. Finally, Mary said, "Do a couple of you guys want the oars?"

That sparked a nervous round of laughter before Mark and Steve took the oars.

Ron said, "If we would have got swamped, we would have blamed it on them being women drivers, but you guys will have no excuse. Holey mother of us, that's enough with the jokes. Here this thing comes."

The wind had picked up, and it hit them directly in the face. Mark and Steve rowed at a fast pace and headed straight for it. When it was close enough, they could see it was a swell carrying what looked to be saltwater foam. There was so much of it the waves looked bigger than they were. The boat went up and over a wave and right through the foam. They all ducked, but there was foam on top of them and in the boat. They all agreed they had never seen anything like it before as they moved their arms about to remove the foam from the tops of their head and faces. They had seen foam on the beach after huge swells, but they were so far out to sea that it was eerie it was there. They were making their way through the waves and gradually slowed the pace down to hardly moving. The current was against them, and they could barely make any progress. They were all looking around at what appeared to be foam in every direction. Every once in awhile, the boat would rise, and as far as they could see, there was foam everywhere. They were in awe of the view but getting wet, so they decided to cover the boat to get under the tarp and wait it out.

The boat was rocking, but it was gentle enough that they didn't worry about being tipped over. They talked about how "freaky" that foam was and how none of them had seen anything like it before. It wasn't like they had all lived their lives out at sea, so maybe it was something that had happened before, but they thought it was probably a rare event caused by the previous tsunami. They remained confident they had been going in the right direction. They could now feel the boat moving in a current that was taking them in the wrong direction.

Ron said, "Oh well, easy come, easy go, and we'll get going again as soon as this passes."

They slowly made themselves as comfortable as they could and tried to sleep. Every hour or so, someone would wake up and lift up a bit of the tarp to have a look and declare it was still foamy. The end of the day arrived, and everyone was taking turns squeezing out of the tarp at the back of the boat. They all drank a cup of water before finding their spots to try and sleep. Ron thought one positive thing was they had all only drunk two cups of water that day instead of the three he'd thought they would. If they kept that up, the water would last them twice as long.

He thought, *Oh God, please don't let us still be out here six days from now without rain. Never mind the rain; just don't let us be out here six days from now.* Then he thought, *Whatever will be, will be. Even if there is a God, there's nothing we can do about it. Either we survive or we don't, so just go to sleep.* Ron called out for everybody to try and sleep and be ready to get moving again in the morning. They all exchanged wishes for pleasant dreams and a good night's sleep before resting in silence for the rest of the night.

Next morning, Ron opened his eyes to darkness again. He hadn't slept through the whole night, but he had managed a few very good patches of three or four hours. He figured he was getting more than eight hours of sleep during a day even if it was an hour here and there eight times a day. With the sleep from the previous afternoon and early night, he must have slept for ten or twelve hours, and some of those had been deep and useful sleep, not like when they were getting tossed around and he could only sleep for minutes at a time. He thought maybe it would be a good day, and he hoped the weird foam would be gone.

There was no reason to hurry, so he rolled over slowly. All that long and useful sleep was great but made his back and entire body stiff and tight, and the pain worse. It was shooting up his back, his neck, and to the back of his head. His feet, his legs, and his toes were aching too as he untied the tarp and stood up. It was still dark, but he could see the foam was gone, and he thought that was a sign it was going to be a good day. He started his stretching routine, arching his back and reaching for the sky. He wondered how far the current had taken them. It could have been further than from where they'd started or not very far at all; he had no way of knowing. He was pretty sure it had taken them the opposite direction but thought maybe they were due for some more luck and they could get caught in a current taking them in the right direction. He thought that with all the luck that had kept them alive so far, if there was such a thing as fate, they must have been meant to survive. Whether it was so or not, the only rational thing to do was to start moving again.

He finished his stretches and waited for the others to start moving. He relaxed and enjoyed the sunrise. He untied more of the tarp and let the light in. The others started moving about, and he uncovered most of the boat. He started rowing while the others were still waking and going through their morning routines. The swell was slowing him down, but he was still able to make progress and get in a rhythm. Steve soon joined him, and they rowed for a few hours.

Finally, Ron said, "Who else wants to row? I'm going to get in the water and push for awhile."

Bob and Mark took the oars while Ron and Steve were in the water pushing the boat and kicking. Every hour or more, someone else would take a turn with the oars or in the water. Ron took a few turns in the water, and he made sure he was in the water at the end of the day. The swimming was what he needed to ease his back pain. When he climbed back into the boat, he would feel great for an hour or two, but then the pain would return. When he jumped into the water, he could feel instant improvement and would be feeling great again within minutes.

When he climbed back into the boat for the last time that day, it was dark. They all had a cup of water, covered the boat, and tied the tarp down before trying to sleep another night.

The next morning, Ron opened his eyes and thought he should feel good after all the swimming the day before and starting the night in good shape. He rolled over to his stomach, and as soon as he started to get up, he knew that wasn't so. The day went the same as the previous. The swell was a little bigger than the day before, and that slowed them down, but they had managed to keep moving all day. They each had only drunk one cup of water at the start of the day and another at the end of the day.

Ron said, "We have about two or three cups of water each before we run out. We've been making good time, so hopefully, we hit land tomorrow or the next day, or else we need rain."

They agreed they would continue with the one cup in the morning and one at night until they ran out. The next morning, Ron opened his eyes and hoped for a better day than the previous, but it was much the same as the day before. He watched the sunrise while he stretched and moaned. They all had turns swimming and with the oars. The children fished with a lure on either side at the front of the boat. They used just two meters of line so they wouldn't get tangled up in the oars. They hoped the lures looked like little bait fish following a boat. They fished for a few hours, then rested a few, and then fished again. At the end of the day, they hadn't caught anything or had a bite. Ron thought they were having fun and feeling good about themselves because they were making a contribution, so it was at least a small improvement for them from the day before. The fishing also kept them entertained all day and helped the time pass.

They had all drunk one cup of water in the morning. Through the day, Ron noticed how dry his throat was. It was causing him some stress, especially after he had been swimming or while he had the oars. It had taken a huge effort to create any moisture in the throat and to regain enough strength to breathe easily again. When he lined up for the nightly cup of water that Steve and Mark were about to start pouring, he asked how much water was there. Steve estimated plenty for one cup each, but not two. They talked about whether it would be better to hydrate more now or stretch it out. They decided and voted on options to have the night's cup a half-cup at a time so they could be sure how much water they had left. Perhaps they would all have just slightly less than a half-cup at a time until they ran out.

When Ron drank his water, it hit the back of his throat and caused a new but welcome pain. He got the rest of it down a little at a time. It hurt but felt so good. He knew it was saving his life. He thought about saving his second half-cup for later, but he was already so dry again he knew he would drink the other half. Ten minutes later, he was asked if he wanted the other half cup, and he lined straight up and drank it slowly. He washed it around with his tongue and noticed how much better it felt after the first drink had moistened the inside of his mouth. He found his sleeping spot and closed his eyes for the night. As he tried to sleep, he assessed the situation and considered it almost a carbon copy of the day before. There was a little less water than the day before, and it was another day without food. At least they were making progress in the right direction, and that had to be the main thing. They had been moving two full days in the right direction, and that was a win. He thought they must be getting close to land, and perhaps the next day would be the day they could put their feet on the ground.

The next morning, Ron opened his eyes and looked at the underside of the tarp. He was surprised he could see light. He had slept past sunrise, and it appeared they all had. As he moved his leg, he felt the pain in his spine and was temporarily paralyzed. He realized he hadn't even gotten a chance to turn over before the pain started, so it would probably be worse than the day before. It took him a considerable effort to turn over, crouch to untie the tarp, and stand up. At least the day was starting differently than the day before, even if it was worse. This would be no carbon copy of any previous day. He stretched and went through what was now his morning routine. As he did, the others woke, and all did much the same. Ron thought group morale was actually pretty good even if his back was worse. They weren't all joking and smiling like before, but they were taking everything in stride in a rather dire situation.

They were all thirsty, hungry, and tired, and some of the others probably had pain, too. Charlie sure did; he had improved, but he was still tender and tired easily. They really couldn't be expected to be deliriously happy, and anything short of giving up or freaking out had to be a good thing. His throat was just as dry as it had been the day before. They had their morning half-cup of water, with Charlie and Sue pouring. They declared there might not be another half-cup each for the nightly drink. They all enjoyed what they had ,washing it around the insides of their mouths before swallowing a little at a time.

Ron and Bob took first shift with the oars. Ron was hoping to time it so he would swim in the morning, in the middle of the day when it was hot, and be the last one swimming at night, or that they would find land. Finding land or being spotted by an airplane would be the best outcome. Ron wondered why they hadn't seen a plane while the days had been so clear and guessed it was because there weren't many flying with all the earthquakes, volcanoes and tsunamis. He went through the day using motions and gestures and speaking very little to be easy on his dry throat. They all had been quiet and said little when changing from the oars to swimming or resting. The sun was getting close to setting when they could see something in front of it on the horizon.

"Yeah it's the rubbish pile,' said Steve."Let's go onward to Rubbish Island."

It was still so far away that it was only when the boat would rise on a wave that they could see it clearly. It appeared there were a lot of active birds. Ron encouraged the kids to get their lures in the water. They had fished without luck a few times during the day, but Ron excited them with the notion the fish would be more likely to bite when they were getting near the rubbish pile, especially with birds working at sundown.

Ron volunteered to get in the water. He wanted to take the weight off his spine, but more so, he wanted to push the boat as fast as he could. They pushed the boat as fast as they could and kept the pace going for almost an hour before slowing down as the last ray of light disappeared over the horizon. Just then, little Eddie had a strike on his hand line. The fish jumped out of the water and dropped back in again. Mark had an oar and leaned towards the front of the boat. As Eddie tried to pull the fish into the boat, Mark put the oar under it and lifted it before flinging it into the middle of the boat. The fish was a tailor, and it was flipping madly until Jin put his hand over it and then used his knife to cut under its throat. It was sudden excitement that put a smile on everyone's face.

The next big decision was how to eat it and when to have their swallows of water. They decided to eat it raw and ration the water out one bottle cap at a time. The cap from the top of the container served as a measure. They decided to have a little at a time so they would all get some and then, when they ran out of water, all anyone would have missed was be that little amount. They all had a cap of water before they started to prepare the fish. Jin had been holding it over the side, letting it bleed out of the boat. He cut off the head and asked who wanted it. Nobody spoke up, so he handed it to Julie. She took it, looking rather uncertain as to what to do with it, and started picking little bits of meat with her fingers and eating it. He scaled the rest of it over the side of the boat and used a little of the seawater to rinse off the last remaining scales. He asked who wanted the guts, and nobody spoke up, so he cut open the fish's stomach and put the guts into Eddie's and Jessie's cups. He said Eddie should have them because he had caught the fish, and everyone agreed. Jin and Eddie looked so serious, and everyone else was smiling at the notion Eddie wanted the guts.

When Jin started handing out strips of flesh, everyone went quiet. The only sounds were groans of approval and requests for salt. When they were finished, Ron guessed everyone had been able to eat several strips, about a mouthful each of raw fish. That was a reasonable meal, especially when they'd been expecting nothing. They all had another twocaps of water before they declared that water was gone.

Ron closed his eyes that night thinking that at that particular moment, they were ok. It was the next day they needed to worry about. They were out of food and water, but hopefully, they would get near the rubbish pile and maybe catch a fish or snag a bird, and it had to rain sooner or later. He tried to think back to the last time it had rained and count how many days all that water they'd collected had lasted them. He tried to count the days they had been out there, stuck on the boat, and realized he had lost track.

The next morning, he opened his eyes. He had spent most of the night awake with his eyes closed and had only managed a deep sleep in the hours before he woke. He was excited for the day to begin and hoped they would be able to see Rubbish Island as soon as the sun rose. He struggled with his stiff muscles to untie the tarp and squeeze out. He felt better than the day before, and this day would have purpose. The others woke and started moving around. It was calm enough that they felt comfortable removing the tarp in the dark. They left it at the front of the boat, ready to roll out in a hurry if they needed it. They went through their own morning rituals before the sun cracked the horizon. They looked in the other direction, hoping to see their target. It wasn't visible, but they knew the direction they wanted to go, so they started moving.

Ron and Steve were first in the water, and Bob and Mark took the oars. They had been moving at a steady pace for almost an hour when someone from the front of the boat shouted out that they could see something ahead. Ron and Steve couldn't see from their position in the water, but everyone else concurred they could see Rubbish Island. Bob and Mark slowed to have a look but kept them moving. Ron and Steve asked to keep going while they were making time. They kept the pace for another hour before Steve called for a break and Bob swapped places with him. Ron went for another hour and called for a break. When he climbed into the boat, he had a good look at Rubbish Island. It looked to be just as far as it had been when he'd seen it the day before. He wondered how that could be after they had made such good progress. He thought it best to keep his disappointment silent and just keep the boat moving. Mary and Amy had the oars, and Deb and Sue were in the water, so Ron made his way towards the front of the boat and found a place to spread out.

He managed to fall asleep for a few hours. When he woke, he tried to bring up some moisture from his stomach to his dry throat as he looked at the sky and saw it had clouded over. He felt a drop of rain on his forehead and then another on his ear, and he popped his head up suddenly.

"It's going to rain," he said. "Get something to catch the rainwater."

He rolled over and got up and then found his coffee cup. He looked up at the sky and called for everyone to get ready. Then, as if he were taking his own cue, he grabbed the water container and made his way to the front of the boat. He called for help to use the tarp to funnel rainwater into the container. It started raining, slowly at first, and then it poured big drops for a few minutes. Then it slowed for a few minutes and stopped. They had all filled their cups and anything else they had that could hold water. The water container was almost half full. They had all been holding their mouths open while it was raining, and they drank their cups after it stopped. They wrung their shirts out over their heads and the backs of the necks to rinse off the salt and dirt, and then they used them to mop up the floor.

The rain had stopped, but the clouds still covered the sky. They couldn't see the sun but could see the rubbish pile, so they knew which direction to go. Ron took an oar, and Bob joined him. It seemed nobody wanted to get into the water now that the wind had picked up and the temperature had dropped. The swell had grown, too, and soon, they weren't rowing as much to go forward but to avoid capsizing. Ron and Bob held the oars the rest of the day.

They commented on how fast the day flew by when they had to try to keep the boat from tipping over. They couldn't see very far or see Rubbish Island any longer, and they put it down to the clouds and poor visibility. They were unsure which way they were going. Riding the waves up and down, and with all the wind throughout the day, had them all turned around. They agreed it was time to stop for the night and try to sleep. They made the boat as watertight as they could and settled for another night under the tarp in a wildly rocking boat.

The next morning, they woke to the sound of rain and the motion of the boat rocking. Ron hadn't slept well most of the night, but he had been sleeping deeply before he'd heard the rain. He wasn't sure what he was hearing. He couldn't open his eyes. He was dreaming and trying to wake up but was somehow paralyzed. Was he just dreaming it was raining, or was it really raining? All of a sudden, he popped up.

"It's raining again," he said. "Get your cups." He rolled over and crawled to his cup and then to the back of the boat and said, "Can one of you bring the water container with you?"

Ron moved as fast as he could, which was rather slow due to his stiff muscles and joints. He was at the back of the boat and outside of the tarp, filling his cup with rainwater before he tried to stand. His cup was full, and he started to drink as he struggled to straighten his back.

Bob had the water container with him, and he set it at the end of the boat and held the tarp to funnel the rainwater into it. The water container was half full when they started, and five minutes of the steady rain filled the rest of it. The rain stayed all day. When it would let up, they would take the tarp off the back and relieve themselves. The little bit of raw fish had given them all a small case of the splats. They laughed it off, saying it was worth the trouble and that they would like to eat more raw fish even if it did give them the runs. They joked that if fish were attracted to their waste, maybe the splats were a good thing, and they should have a lure in the water. When the rain poured afterward, the sound helped them get to sleep. At the end of the day, they had merely stayed out of the rain and made no progress. Ron thought at least they had drinking water. They had been lucky again. The rain had saved their lives, and they would survive a while longer.

The next morning, Ron was flat on his back with his eyes open as he waited for the light. It had stopped raining, and the boat was rocking gently. He struggled to turn over and get up, and he was happy that it seemed easier to manage than most of the previous mornings. He stretched his spine as he watched the sunrise. It was cloudy in patches and windy, so when he couldn't see Rubbish Island, he thought maybe it was just poor visibility. He could see the sun in between the moving clouds, and he knew which way they wanted to go, so he volunteered to jump into the water first, and they started moving. A few hours later, he climbed back into the boat and took one of the oars. He rowed for a few hours and figured it must be around high noon. They still couldn't see Rubbish Island. Was it behind them because they had passed it while they'd drifted at night, or was it still ahead? It was clear and flat enough to see in all directions, and there was no sign of anything in sight.

Ron rested for a few hours before having another look around and getting back into the water. He pushed the boat for another three hours before climbing out of the water. When he had caught his breath and was sitting on the bench seat, he said, "I like it when my fingers and toes go white like this; they must be really clean."

That started conversation and jokes about wanting soap and a long, hot shower and how much water they would use once they could have as much as they wanted. Having water would be great when they only had to turn a tap. They all drank as much water as they wanted during the day and another cup before they started to cover the boat for the night. Bob was the last one to pour a cup of water, and when Ron asked how much water was in the container and Bob said it was three-quarters full, there were groans and nods of approval.

The next morning, Ron woke and repeated his routine. He still couldn't see Rubbish Island. It was a windy, cloudy day, and the choppy water was rocking the boat up and down, making it difficult to see far. The others seemed a little slow to get up and about, so Ron took the oars and started rowing on his own. Only the back half of the boat was uncovered as Ron, Bob, Steve, and Mark took turns with the oars for the first half of the day.

Nobody had gone for a swim. Ron said he thought it helped to have someone in the water pushing, and he tied a rope around himself and to the boat and jumped in before Bob joined him.

Charlie and Sue took the oars, and they found a steady rhythm for a few hours.

Ron climbed back into the boat, elated he had helped his aching back and made a contribution to their cause. He caught his breath while his eyes scanned the horizon.

He said, "There it is. There's Rubbish Island. It's further than it was two days ago. What is going on? Are we moving backward, or is it moving?" There was silence, and he added, "Either way, all we can do is to keep going."

Mary and Amy jumped in the water to have a swim, and Jin and Julie took the oars. They found a rhythm and kept it steady until dark. When they couldn't see where they were going, they stopped and made ready for another night.

The next morning was much the same as the others. Ron struggled with his back pain and to straighten his body. He looked out in all directions as the sun came up, and he could see the rubbish pile. It looked like it was still a long way away, but the fact that he could see it so early meant that it had to be closer than it had been the day before. He made his way to the water container and poured a cup of water as the others started to wake.

"I can see Rubbish Island straight off the bat this morning," he said, "so I'm keen to get going. I'll go for the first dip this morning."

They took turns with the oars and in the water like they had on previous days. Each hour or more, they would switch around and comment on the progress they were making. They weren't sure if they would reach the pile before dark, but they thought they would at least get close.

Ron got the kids excited about fishing again, and they had lures on both sides of the boat. They were watching the sunset over the rubbish pile when little Eddie got a strike on his lure. He pulled in a reasonable-size tailor, about a foot long. As he took it off the hook, his sister hooked one and pulled it straight into the boat. Eddie threw his lure back out. He let it settle and then moved it a little with a jigging motion and hooked another. They had caught nine between them before they decided to cover the boat and call it a night. Then Jin bled, scaled, and gutted them. The kids were eating some of the guts raw and had saved some for bait. Ron wondered if that would eventually make them sick or if they were used to it, or if it was because nobody ever told them anything different that they liked it and didn't vomit. They put the scaled fish with gills and guts removed into the foot bucket and strapped it tight with two hockey straps to hold it tight overnight.

The next morning, Ron woke much the same as before. He thought, *Oh crap*, but then he remembered they had fish and water and were close to the rubbish pile. He made his way to the back of the boat and squeezed out of the tarp. It was cloudy and cooler than the previous morning, reasonably calm, and the wind was light. He could see the rubbish pile, and again, it was further away than it had been the night before. It was at least closer than it had been the previous morning, but it was strange it was so much further than the last time he had looked before getting under the tarp. It was a cool gray day, and maybe they could reach Rubbish Island with just the oars. He felt his back was lousy, so he wanted to go for a swim, but later, when it might be warmer.

He went through his morning routine and was rowing the boat when the others woke and started going through their own morning routines. He rowed for a few hours and swam for a few hours and then took a nap. He woke again and noticed how hungry he was and figured the others must be hungry, too. He looked out and saw they were getting closer to Rubbish Island. If they stopped to eat, they might not reach the island.

He said, "Has anyone thought about when we're going to eat those fish?"

Steve answered, "We were going to see how close we can get before dark and eat then."

Ron nodded and said ok. During the day, he took another turn in the water and two with the oars. Just before sunset, they were approaching Rubbish Island and were close enough to keep moving in the dark and know they would reach it.

They spotted a wooden stake floating near the edge of the pile and took the boat through the plastic bottles and other floating debris to grab it. The stake wouldn't be enough timber to make a fire big enough to cook the fish, so they kept looking for more. When they couldn't find any, they decided to eat the fish raw. They could save the stake, and if they found another one, they would have enough to cook the fish next time they had some. They ate the raw fish after cutting it into small strips like they had before. It was chewy, but the pieces were small enough they could swallow them whole. They had seen a few sharks swim past, and there were birds all over the pile, so they were confident they could catch something to eat the next day. It was dark when they were covering the boat, but they had become used to the routine and tightened down the tarp by feeling their way around the boat. They gathered what they had to make themselves comfortable and found their spots to try and sleep another night.

The next morning, Ron woke and was quick to get started. The boat was rocking gently as he pushed his body through the small opening in the tarp. It was still dark as he stretched and tried to straighten his back. He opened up more of the boat as the others started to wake and move around. When there was enough light, they could see the rubbish pile. They had drifted again, but this time, it was much closer, and they knew they would reach it easily. Ron and Steve took the oars, and within an hour, they were at the edge of Rubbish Island. Their plan was to circumnavigate it while looking for bits of timber or anything to burn, and at the same time to catch fish. When they were on the other side, hopefully, they would have something to burn and something to eat and would be close to land.

They spotted a section of a timber fence maybe twenty meters inside the edge. They hadn't gone that far into the rubbish before. They had worried they could get stuck and not be able to get out, but the amount of timber looked to be worth the risk. They pushed the boat through the rubbish and grabbed the fence. It was six feet by four feet and rather heavy. It appeared it was floating on enough plastic bottles and other rubbish to keep it afloat, or it might have sunk. It was too big to bring into the boat, so they hung on to it as they pushed the boat back through the rubbish and outside the edge. They used what tools they had and managed to take it apart and pile the timber into the boat and tie it down with hockey straps. They were happy with their find and then were keen to catch more fish and have the wood to cook them.

They slowly made their way around the pile as they fished. Then they stopped for awhile so they could fish deeper. They had been fishing for hours before they caught their first fish of the day. It was a very small tailor, so small they decided to cut it up and use it for bait. They caught a few more small tailors and thought that with a few more of those, at least it would be a little feed. They caught four more, and as soon as the bait sankto the bottom, it was taken. The next tailor caught was another small fish, and Ron put a hook through its tail and dropped it back in alive. Several minutes later, he had a huge strike, and several minutes after that, he pulled in a shark around four feet long. When he pulled it into the boat, all the women went to the front of the boat. It had been a strange fight getting it to the top, and now it was in the boat, still trying to bite anything it could touch. It thrashed around, chomping its teeth and trying to attack. Jin put a large knife through the top of its head and held it up in the air as it thrashed for another minute or two.

"Now we have a feed," said Ron.

Bob nodded. "It's getting dark."

"Yeah, let's get the fire started."

Ron splintered the wood into small shavings, and Bob dug out the portable barbeque. Jin, with Julie and the kids, scaled the tailor and then removed the gills and gutted it. Mary had dug out the pot and filled it halfway with water. The kids dropped the tailor with the heads attached into the pot. Jin cut and gutted the shark from under its gills, taking off the head and leaving fillets almost three feet long. He removed the skin with pliers and cut the flesh into pieces about one inch thick, three inches wide, and four inches long. There were more than twenty pieces dropped into the pot. They kept the head of the shark and all the fish guts in the foot bucket. They would use that for bait in the morning if it didn't get tipped everywhere during the night.

Within ten minutes, the fire was started. Bob picked up the pot and set it on top of the barbeque plate. He and Ron took turns stoking the fire, and within five minutes, the water was boiling. They stopped adding fuel to the fire and were pleased with how little timber they had used. Now their problem was that with this pile of wood and pot and the fire, there was less room in the boat. They were cramped in and kneeling on whatever was being stowed. The fish simmered for several minutes.

Ron and Bob served the fish with a few shakes of salt and passed the plate or cup back down the line. Everyone had at least two servings, and it seemed a lot compared to what they had been eating. They all commented on how much they liked the shark. It didn't have any bones and was really easy to eat. They knew shark might have heavy metals like mercury but decided it was worth the risk. They made jokes about heavy metal music and how it might have something to do with the fans eating fish with heavy metals that made them brain dead.

Ron said, "That's why they fall into a trance, stomping and jerking their heads around to that awful noise they call music."

They made jokes about eating smaller sharks only, because smaller sharks might have less heavy metal than the larger ones, and they didn't want to see any big sharks anyway. The bigger sharks might eat them instead.

They all had a cup of water and agreed they felt pretty good. They would go around Rubbish Island, catching fish along the way, and the next day, they would be on the other side. Then it was probably only another day until they were on the mainland, or at least they hoped. It was dark, but calm and not raining, so they decided to sleep under the stars and left most of the boat uncovered.

The next morning, Ron woke up and thought things looked very different. He was looking at the sky instead of the underside of the tarp, and he wouldn't have to squat while he untied it. It should be a positive start to the day and a benefit to the pain in the back from the start. There was light in the sky, but it was still going to be a while before the sun would come up, so he decided to stay still and relax his muscles as best as he could before he had to get started. He took a few deep breaths and let them out and rolled over to his stomach. Before he stood up, he waited for his pulse to slow down. He took a few more deep breaths and let them out before he stood. His back had made some noises and was still sore and stiff, but it was so much better. He smiled and thought it would be a good day.

He thought back to how the day's routines were becoming the same as the day before, but now there was more hope, and they had eaten and were all feeling good. He relieved himself and washed his face and hands in the saltwater before crawling over to the water container and pouring a cup of water. The rest of them started to wake as he was standing and looking over the rubbish pile. He couldn't see anything in the vicinity. They didn't have any use for more plastic bottles, and the further he looked into the pile, the less he could see what things were. Everything in the distance looked like it was covered with puke or suds or sludge or something. He thought maybe it was just seafoam and strange shapes with colors behind them, but it was better to merely wonder than try to get the boat in there for a closer look.

Ron and Steve took the oars, and the children, under Jin and Julie's supervision, started the fishing. They would row for a few minutes and then stop while the kids dropped a baited hook and the rest of them eyed what was in the rubbish pile in front of them. The sun popped up and went straight behind a cloud. They had been moving for ten or twenty minutes when little Eddie caught a fish. Everyone stopped to watch, and they quietly cheered as he was bringing the fish in hand over hand. They could tell it was giving a fight or was a heavy fish. As he brought it to the surface, Jin reached into the water and grabbed it with his fingers by the gills. It was a flathead around three feet long, which was large enough to feed everyone a reasonable meal. They commented how good it was to have a flathead, just a good, heavy fish, not a heavy metal fish. They all said they liked the way flathead tasted and shared the different ways they liked to cook it. They didn't have any garlic or lemon, but all agreed flathead would taste great as it was.

Jessie hooked another fish, and excitement started all over again. It seemed a smaller fish at first, but then he gave the line a mighty tug. Her mom was helping her and caught the spool when she dropped it briefly, but she was bringing it in hand over hand and straight out of the water and into the boat. It was another flathead. This one was only two feet long, but it was still a nice fish. Jessie had a huge smile on her face and was bouncing with excitement. Jin was taking it off the hook when Eddie hooked another one. He pulled it in hand over hand without fuss. Jin had bled and removed the gills of Jessie's flathead as Eddie pulled it into the boat. It was a little smaller than his sister's.

"They're supposed to get bigger," said Steve."Keep trying."

They fished for another hour without a bite. The three flatheads had been scaled gutted and had their gills removed. The fish heads were still attached and were in the foot bucket. The guts were squeezed into a plastic bottle, and the lid screwed on. Now they had two bottles of fish guts for bait or burly, but because they were about to start moving, it was decided to fish with lures and trawl shallow while they increased pace. Ron and Bob went into the water while Steve and Mark took a turn with the oars. Ron was feeling great in the water, and they were making fast time. They went steady for two hours straight. When Ron climbed back into the boat, he had a look around and shook his head with a puzzled look on his face.

"There isn't any sun," he said. "How do we know when we're on the other side and want to head towards land, or will we just keep going around Rubbish Island in a circle?"

They stopped rowing and thought about it. It hadn't felt like they had changed directions, so they agreed they would stay close to the pile until the sun came out and they could be sure which direction they were going. They sat around ,but the only thing they could see was the pile of rubbish.

"What is that?"asked Ron. "It looks like part of a ship sticking out of the water. There's a steel grid floor that looks like an upper level of a container ship."

"It's not very far in," said Bob.

"Let's go have a look at it."

They pushed a path through the plastic bottles and bits of floating rubbish. There was a large plastic storage box with a lid among the rubbish next to the boat, and Ron reached down and grabbed it. He pulled it into the boat, opened it up, and cheered loudly as he pulled out a spool of steel wire and a pair of pliers.

He said, "This will be useful, and there are some tools, a hammer, screwdrivers, a chipping hammer, a striker, a few files, a steel brush, and safety goggles. There's some string and a tape measure but nothing much worth anything to us. The box has a lid, and it floats, so it will be a good thing to have."

They made their way up to the ship, and Ron grabbed onto the ships handrail. The ship's sleeping quarters and the structure holding the steering wheel were sticking out of the water at a forty-five-degree angle. Ron took the boat to the spot on the ship with the steel mesh floor and pulled up next to it. He stared at the mesh for a moment and declared it was the perfect place to cook their fish. They tied the front and back of the boat to the ship's handrails so it wouldn't move. Ron suggested they cook and eat there, and they all agreed.

He used the steel wire they had found to wrap around each fish, and on the end, he fashioned a foot-long handle. He got out of the boat and onto the ship and prepared to start a fire in an area where the steel mesh floor met with a steel kickboard under the handrail. Bob handed him a lighter with some wood shavings and splinters of wood, and Ron started the fire. When it was blazing, Bob handed Ron the fish wrapped in wire, and he set them in the fire.

They watched as the wire turned red with the heat. Ron looked at the flesh of the fish as it cooked. He wasn't really sure what he was looking for, but after a few minutes, he had a feeling it was time to turn them over. He took off his shirt to handle the hot wire handle and quickly turned over the fish, starting with the smallest one first. He watched them cook for another minute and then called for the pot to put them in. Bob handed him the pot, and Ron, using his shirt and acting quickly so he didn't burn himself, put the cooked fish into the pot. The fire was reduced to ash, but it was still very hot until the fish had cooked. Ron was pleased; he hadn't used much fuel, and it hadn't taken long to cook.

They all enjoyed the fish, and when they were done, they commented on how full they were.

"We didn't use much wood or water either," said Ron.

"Speaking of water," said Steve, "look ahead." There was a storm coming.

"This is good we're here. We can stay tied up to this ship, and we won't be capsized."

They all agreed and tied the boat to the ship as tight as they could, and then they covered the boat with the tarp tied down tightly. Ron had carried the water container to the back of the boat to top it up and keep it full while they could. The container was refilled in less than a minute, and the rain poured hard for hours.

Ron woke up the next morning, and before he moved, he could feel his pain was worse than before. He has slept in patches after being woken up every time a wave had hit the boat and smacked it up against the ship. Being tied tight up to the ship had stopped them from rocking around or capsizing, but hadn't helped them sleep easier. It took a few attempts before he could roll over to his stomach, and it was slow going to untie the tarp and get himself through the small gap. It was steady rain, so he had a look around the boat and ducked back under the tarp and tied it down again. He found the empty plastic bottle he had picked up and saved the night before just for the occasion. He peed into the bottle and then untied the tarp so he could squeeze out and empty it. He rinsed the bottle and dropped it on the floor before taking off his shirt to mop up the rainwater that had come into the boat.

When he was done, Mark and Steve crawled to the back of the boat. They decided everyone would use the bottle and mop the floor when they were done. The girls could use the foot bucket or wait for the rain to let up. If anyone needed to sit over the side, they should wait for the rain to let up, and if they couldn't wait, they would need to hurry. Ron said that when he'd looked outside, he hadn't seen anything exciting. There was too much rain to see very far, but he could see the sea was very choppy. They decided they were better off where they were until it stopped raining and the sea calmed down.

They stayed put all day, and the only action was when someone needed to pee or sit over the side, which was particularly uncomfortable due to their fish-only diet. They agreed they needed some vegetables or something with fiber, but in the meantime, if fish was all they had to eat, they would cope with it. They all agreed they should drink plenty of water while they could so they wouldn't get dehydrated.

The rain stopped briefly late in the afternoon, but they decided it was too close to dark and the seas too rough to set out. Half an hour later, it was pouring again, and they commented on how they had made the right choice. They were all hungry and ate what fish was remaining raw when they decided they couldn't wait another day to hope it stopped raining so they could start a fire and cook. They expressed both hope and confidence they would be able to catch more fish as they passed around the raw fillets. Straight afterward, they found their spots and tried to sleep. Ron found it particularly difficult that night with his back stiffened up and other muscle spasms and cramps.

Ron woke up sore the next morning and thought he would stay put and wait for someone else to get up and untie the tarp. It had stopped raining during the night, but the seas were still rough and slapping the boat. His pain was more severe, and he knew he needed to get up and move around, but the thought of crouching to untie the tarp swayed him to wait for one of the others to wake up first. He had been the first one up just about every other morning, so he would let one of the others get them started. He knew they knew he was having back pain, but he didn't want to complain or let them know how much pain he was in. He didn't want them to worry about him or think he was looking for special considerations, or for them to look after him.

He remained still for an hour with his eyes open and his mind rambling. He knew someone else must be awake. They couldn't all be asleep with the boat banging into the ship every few minutes. He wondered if anyone else was going to get up or if they were all waiting for him to get up and open the tarp. He thought he had better get up so, if the conditions allowed and they could get moving, they could do it faster. He struggled to roll over and get started. When he opened the tarp, the others started to move around.

He stood there making a considerable effort not to scream or yelp as he straightened his spine as best as he could. The others uncovered the boat while he stood there hanging on to the back of the boat with both hands so he wouldn't fall when a wave knocked them about. They all went through their morning routines as Ron stood there not making a sound. They decided it was worth going ahead with the oars even though the sea was somewhat rough. It was too rough for anyone to be in the water, and if they were in the water in those conditions, they would only slow the boat down. They didn't want to stay put in the same place for another day without food or making any progress towards land. The kids were keen to fish, but the adults doubted any fish would bite in the rough seas.

They untied the boat, and Mark and Bob took the oars first. Every hour, they swapped seats, and someone else took the oars. They were going up and down and managing to keep the boat upright but were making little progress. It was cloudy, but there was enough sun getting through for them to see where they were going. The kids fished off and on all day, mostly with lures as they trawled, but they didn't get a single bite. At sunset, they were still looking over Rubbish Island and needing to get around to the other side. Ron was happy they had made progress, but the others weren't enthused because of how little they had made. They all agreed they were happy they at least hadn't gone backward. As the sunset, Bob and Mark took the hand lines from the kids and baited the hooks with fish gut, but to no avail. They would have to sleep another night without food. There was no ship sticking out of the rubbish pile now, or anything else they could see to tie onto, so they would be drifting all night again.

The next morning, Ron woke with the same pain as the day before. He was curious to see how far they had drifted from the rubbish pile, or if they had drifted into it. Getting the boat stuck on top of Rubbish Island would be a problem they hadn't had. It was still dark, and he stretched and kept his arms and legs moving while he waited for the sun to be high enough to see the horizon. He could see they had drifted away from Rubbish Island again. He could see what he believed was the spot they had spent the night tied to the ship and the spot he thought they had reached before the previous night's sunset. It was a fair distance between those two points but probably not as far as they had drifted during the night. One positive thing he saw was the end of the pile and how they could steer the boat in the direction to where they could get around it and start heading for land. The sea was calmer than the day before, and they were happy to be able to make better time.

The others were uncovering the rest of the boat, and Ron pointed to the corner of the pile he suggested they should head to. He pointed to the spots where he thought it looked like they had been and where they needed to go, and the others agreed. He was the first one in the water and waiting for the others to get situated. Bob jumped into the water, and when Mark and Steve took the oars, Ron and Bob started kicking hard. They hadn't started rowing, and the swimmers were already pushing the boat along. They started slowly and built up speed to a steady pace they could maintain. Eddie and Jessie were trawling with lures on either side of the boat, and they kept the boat moving for a few hours without stopping.

When they stopped, Bob and Ron climbed back into the boat and were physically spent. Ron was tired, but his back felt so much better that he had a huge smile on his face. He found a spot to flatten his back on the floor and got as comfortable as he could while Mary and Amy took the oars and Sue and Deb jumped into the water. The boat started moving again, and Ron fell asleep. He slept until the middle of the afternoon, and then he and Bob took a turn with the oars. They kept a fast, steady pace for a few hours and then took another turn in the water. It was near sunset when they climbed back into the boat. They were still near Rubbish Island and further along than they had been the night before, and they were pretty happy about that. The kids hadn't caught any fish, and Jin and Julie took the hand lines. They left the lures on but added some fish guts to the hooks. They thought their chances were better now the boat had stopped, it was dark, and they could fish deeper. Still, they caught nothing, and eventually, they gave up and covered the boat with the tarp before they all tried to sleep another night.

The next morning, Ron woke up feeling a little better than the days before. The early-morning and late-afternoon swims must have been the dose he needed, and he planned to do that again. He was standing and stretching as it became light, and he could see Rubbish Island. They had drifted away from it again but not very far this time, and they had also drifted towards the end of the pile. He could see they would be rounding it that day and maybe find land the next day. His pain had improved, and they were almost there, and he smiled as wide as he could while he scanned the horizon. Then a sudden hunger pain caused him to flinch and hold on to his gut. *There always has to be some kind of problem*, he thought, but at least they would all be having a meal on land in a day or two.

They were all awake when the sun came up. They watched it rise until it was bright, and they drank cups of water while they discussed how far away from the pile they were, how long it would take them to get there, and how long they should circumnavigate it before veering towards the mainland. They didn't know where they were, and they could end up back in Sydney where they started or anywhere else up or down the coast. They wanted to find the nearest place to land and not end up in Tasmania because they had turned too early. They agreed to just go around it for the day and see where they were at when the sun set.

Ron and Bob took the first turn in the water while Mark and Steve took the oars. They made good pace and were nearing Rubbish Island. They had paddled towards the corner and could see they were only a few hundred meters from rounding it. They would be changing direction, but they couldn't go wrong as long as they kept Rubbish Island to their right. Bob climbed into the boat, and Charlie joined Ron in the water. They were all happy about watching Charlie feel better and helping out, and some claps and encouragement let him know it. They decided they would trawl while they were moving but not stop to fish until the end of the day. They would keep an eye on the rubbish pile but wouldn't stop unless they saw something they needed and could grab it quickly.

They stopped after another hour. They had certainly rounded the corner, but they still couldn't see how far it was to the next corner. They watched the sun and thought they were at least heading towards land and concluded it would be a guess which direction from that point would be nearest to land. They decided to wait until sunset, when they could be more certain of the direction they wanted to head. They took turns swimming and with the oars. Everyone was feeling confident they were making progress and would find land soon, and about an hour before sunset, they knew they were heading directly towards land even though they couldn't see it.

They stopped and fished. They hadn't seen any shark or fish all day. They could see birds further in the rubbish island, but they weren't very active either. There was only a small amount of bait left, and what they had was shriveled and deteriorated. They drifted near the rubbish and paddled now and then to stay close to the pile and keep moving in the right direction. Ron spotted a dead seagull floating on the edge of the pile. He rowed to it and used an oar to dip under it and bring it into the boat.

Bob said, "That's not enough to feed all of us," and that drew laughter.

"We can use it for bait," said Ron.

The fish guts they had left were useless, so they rinsed out the bottle to make burly and then cut off the dead bird's leg and a piece of its wing to bait the hooks. The kids fished until dark. They had trawled almost all day, so it was only fair to let them fish with bait. They were pretty good at it anyway, and with everyone watching intently, no fish would get away. No fish bit, though, so they covered the boat and found their spots to try and sleep.

The next morning, Ron woke up and went through his usual routines, only this time, when he had done his stretches and was waiting for the sunrise, he dropped the hook with the seagull leg over the side. He waited for the sinker to hit the bottom and then took it up a few feet of line and waited. He fished with the line over his finger and gave it a twitch from time to time.

When the others had gone through their morning routines, they decided to get moving. Ron removed the hooks and tied the lures back on. He asked the kids to trawl and said the bait would deteriorate in the moving water but that if they saved it, they could use it when they stopped for the night. Ron hadn't had a bite, and now that they were on the other side of Rubbish Island, they would be heading to land. They thought they should get there within a day, but it would have been nice to catch a fish just in case it was longer.

Ron and Bob took to the water, and Mark and Steve took the oars, and the day went like the previous day, only this time, Rubbish Island became further and further away until it was out of sight altogether. At the end of the day, they were heading for the sunset and could only see the ocean in all directions. The rubbish pile was well behind them and out of sight. Ron changed the lures and handed the hand lines with baited hooks to Eddie and Jessie, and the rest of them watched the children fish until after dark. They didn't get a bite, so they pulled the lines out, and then everyone covered the boat and made their way to their spots. Ron flattened his back on the floor and held his knees up. His back was sore, but the hunger pains were competing for attention. He couldn't remember how many days they had been out at sea, and now he couldn't remember how many days it had been since they'd last eaten. He wasn't certain if it had been three or four days before when they had eaten that wonderful flathead.

They expected to have found land, and they still couldn't even see it. They thought it a blessing that at least they were closer and still alive and making progress. Ron thought back to the freaky things that had happened to them from the time the tsunami had pushed their car off the road. There were already so many strange things that had happened so they could survive, and he believed that after all that, they must have survived for a reason. He felt certain they wouldn't die out at sea, and he turned his mind to think about what they had to do next. It was the same old story: get some sleep while they can and then wake up and start moving in the morning, and sooner or later, they would reach their destination.

The next morning, Ron woke and again struggled to get out from under the tarp and stand up. It was still dark, and it was calm, and he thought that a good start to the morning. He hurried through his stretches and then found the hand line and started fishing. He watched the sunrise, and as it did, he could see Rubbish Island. He thought, *That can't be good*. They had paddled out of sight of it sometime in the middle of the afternoon so they must have drifted a fair way in the wrong direction overnight. It seemed they could drift faster in the wrong direction than they could paddle in the right direction. It would be at least three or four hours' worth of rowing to get back to where they'd been. He thought, *That's what they mean by the saying one step forward and two steps back*. He called the others to wake up or start moving and to have a look at what he'd seen. They all made sounds of disapproval, but they all got started, and the boat was moving within ten minutes.

Ron and Bob swimming and Mark and Steve on the oars was becoming routine, as were the changes throughout the day.

Ron spent as much of the day as he could in the water. They had been moving for three or four hours when the noticed they could no longer see Rubbish Island.

Ron said, "That's twice we've seen the last of that place. Let's hope it really is the last."

They had been paddling for several more hours when they stopped to fish and watch the sunset. It was a beautiful night, with a light breeze and the boat gently rocking. They watched the last bit of sun and then the light disappear. They kept fishing using the rest of the bird for bait. They made jokes about how hungry they were, how much meat could be left on the bird, and how sick could such little meat make them? They decided it wasn't enough meat to be sick about and they might as well go big with the bait. They tried one hook with the feathers attached and the other hook with the raw breast. They were all certain they would eat another bird if they found one, but they were hopeful the bird in their hand, or at least on their hook ,would attract a shark or large mulloway that would be enough to feed them all.

They tried jigging the bait, changing one of the hooks with a lure, and adding feathers and a bit of bird meat on it. The water and the night were so calm they felt confident they could leave the boat mostly uncovered and fish for the entire night. They crowded together at the back of the boat and left three people at the front of the boat enough room to stretch out and have a sleep. They took turns sleeping in three-hour intervals, and they swapped the hand line seats every hour or so. The rest of the time, they remained quiet or whispered so they wouldn't scare away the fish.

Everyone tried their hardest to sleep soundly for their three hours and catch a fish when it was their turn, but they only managed to sleep for the three hours, which didn't seem like enough when they were woken up. They had discussed paddling at night but had differing opinions on which direction to head. They weren't as sure of their directions when looking at the moon instead of the sun, so they agreed to paddle slowly in the most likely direction while they concentrated on catching a fish.

Ron did enjoy his three hours of sleep. He had fallen directly asleep and was in a deep sleep when they woke him. He thought three hours of good sleep might be better than eight hours of intermittent sleep. He was still very hungry, sore, and tired when the sun came up. The wind was stronger, but the seas were mild and the clouds few. Ron and Bob took the first spell in the water again, and Ron felt tremendous joy when the weight was off his spine. He had a huge smile and yelped, and Bob smiled back before they pushed the boat along for two hours. Mark and Steve had the oars, and they all knew they were traveling slower than the days before, but they were happy they to keep a steady pace. Ron and Bob climbed out and took the oars, and Mark and Steve went into the water, and they managed another two hours of steady progress. They all took their turns with the oars and in the water, and at the end of the day, they all had a cup of water. Ron was the last one to have a cup, and as he held up the container, he announced it was a quarter full.

There wasn't enough room for all of them to stretch out. The timber was good to have, but it was taking up room and was uncomfortable to sit on. Ron found a place to flatten his back on the floor and was feeling the pinch. The swim had done him so much good, but sitting with the oars had ruined him again. The pain was shooting up and down his spine as he tried to remain as still as possible. He remained still for two hours, trying to relax and let his mind wander to other things so that he would forget the pain. It was a technique he had used for many years and was only helpful for short periods of time, but it helped him get to sleep that night.

The next morning, he opened his eyes and assessed his pain as he considered what to do next. He could hear some of the others were awake and moving about. He rolled over to stand up and slowly climbed to his feet while hanging on to the side of the boat. As he tried to straighten his back, he thought he saw a bump on the horizon, and he stood as tall as he could to have a better look.

"Look ahead," he said."Do you guys see anything?"

The others had a look and started pointing. It was a long way off, but every now and then, a wave would pass, and they would get a glimpse of something in the water. Jin and Julie started rowing, and Charlie and Sue dropped into the water. They discussed whether it was a boat or something big floating. The shape was kind of odd and not long like a boat. It was in the direction they were heading, so they were about to find out. An hour or so later ,Ron volunteered and went back into the water. Bob and Mary took the oars, and Amy joined Ron in the water, and they started moving with new enthusiasm. They slowed considerably after the first twenty minutes but kept moving for another two hours before Ron and Amy climbed back into the boat.

"We think it's a house," said Bob.

"What?" asked Ron.

"Yeah," said Mary, "a house."

That started speculating if it was floating and how it could have stayed intact.

Ron said, "We'll know when we get there, and my guess is that will be an hour or two. There's not much daylight after that, so we better keep moving."

Mark and Steve took the oars with new vigor while Jin and Julie jumped into the water and the rest of them looked intently ahead at the figure in the water.

Chapter Six
Little House on the Sea

The nearer they got to the house, the clearer it became ,and the more their excitement grew. They would finally crawl out of the boat and put their feet on the ground. There would be other people there, and hopefully, they would have something to eat and be friendly. It was a house high on a rock in what appeared to be the middle of the ocean. They figured it meant they were near land, but there was nothing else in sight on the horizon. They could see a small tree to one side of the house. They could see the branches of larger trees behind the house that weremuch taller than the roof. As they approached in joyous awe of the scene, they looked for any detail to comment on. They could see other greenery around the sides and what looked to be a jetty with two small rowboats below.

Ron thought about how similar the house looked to a unit he had rented long ago when he had lived in Newcastle. It had a large concrete or sandstone front porch painted white. This house was two stories, whereas his Newcastle unit had only been one story with a crawl space for the attic, but other than that, it was the same. His unit in Newcastle had been built for the foreman at Arnott's Biscuits by convicts, with the sandstone porch and window frames, a fireplace, and stained glass windows. This one was much the same, at least from a distance. It had wooden shutters at the upstairs windows and also plant potters where they could see something green growing.

When they were closer, they could see there were several people on the porch and two looking out from the window upstairs. There were two other people on the pier and two small boats with two people not far away. When they were even closer, they could see there were lettuce and cherry tomatoes growing in the upstairs window planters. It was a citrus tree on the left side of the house, and various plants including large tomatoes on the other. The people in the boat were dragging a net around in front of the pier, and it looked like they were handing it to four people in the water, who were dragging it up on the rocks next to the pier.

When everyone in the boat knew they could be seen, they all started waving at the people around the house. The people on the pier and on the porch waved back. Ron and the others were excited and elated. They were finally going to put their feet on the ground, even if it was just a rock out in the ocean. There would be other people to talk to, and they would find out what had been happening and perhaps have some fruit or vegetables to eat. They were all so happy they were almost shaking, and tears were welling in their eyes. It was a feeling they were saved at last. They would get all the latest news, put their feet on the ground, and most importantly, find out how far they were from the mainland.

Steve altered their attitude immediately when he screamed out the word shark in such a way they all recognized fear and panic in his voice. The great white swam towards them and opened its mouth wide as it popped its head above the surface. The mouth looked six feet wide, and almost as tall, and as it gained speed and approached, everyone in the boat was paralyzed with fear. The shark dove under the boat at the last second as if it had meant to scare them.

Ron said, "Ok, let's leave this fish alone."

There was quiet laughter and then silence, and no one was moving. Then either it or another shark approached from the same spot, and then there were two more behind it. The sharks approached the boat and veered away and circled. Ron suggested they keep moving, and he and Bob took the oars and found a gentle, quiet rhythm as several massive white pointers circled and followed the boat.

Ron whispered, "I thought great whites were solitary fish and didn't school.

"Maybe these ones are educated," Bob whispered in reply.

One shark suddenly popped up and bit the back of the boat, its mouth stretched open enough to have its bottom teeth underneath the boat and its upper teeth on top. So much of the boat was in its mouth that Ron wondered if it was attempting to swallow it whole. It had a couple of chomps on the timber as if it were trying it out or tasting it, and then it let go and dove under the water.

Ron whispered, "Yeah, that's tough chewing. Go away and leave us alone."

The people around the house saw the sharks, too, and they dragged the net out of the water and carried it up to the front porch in a hurry. They pulled their boats out of the water and ran up the stairs to the porch. They were waving from the porch in a hurry-up-and-come-over-here motion to the people in the boat and pointing to the sharks behind the boat. A shark would approach the boat, and someone on the porch would put their hands on top of their head or over their eyes. Ron and Bob just kept rowing quietly.

Ron whispered, "We're just rolling along. There's nothing going on over here, nothing a shark would want to eat. We're invisible." Then he sang lightly, "Just rolling along, singing a song, and minding our own business."

Bob started to laugh, and then Ron let out a grunt and a giggle before whispering they should be quiet, and then he started giggling again. They were frightened, but at the same time, he felt it was unlikely a shark would take them from the boat. They were nearing safety, and they were trying to keep things light instead of panicking. They reached the end of the jetty, and everyone remained sitting. The people on the porch stayed where they were except for two who came down to the bottom step. They had motioned to the people in the boat to stop when a shark would approach and then to come on in a hurry when the shark would swim away.

When the boat reached the other end of the jetty, the crew all climbed out onto the pier one or two at a time. The people from the house who came out to meet them told them to stand on the steps. They said the sharks had taken people off the pier, and even off the steps they were currently standing on, but not to worry, though, because it was a low tide and the steps would be safe. Everyone getting off the boat was happy enough to stand on the steps and were smiling and shaking with excitement. The man nearest them introduced himself as Dan, and everyone off the boat called out their names one after another.

Ron said, "We were washed out to sea by the tsunami that hit Sydney."

"That was more than five weeks ago," said Dan.

"Yeah, we weren't sure how long we were out there. Did you see the news reports and if they mentioned people being swept out to sea and hanging on to the Harbour Bridge?"

Steve added, "Or the Opera House sailing in the harbor? Did they have any of that on the news?"

"That's old news," said Dan. "It was pretty depressing for more than a week; it was all about recovering bodies. There were a few dozen soldiers and around two hundred volunteers working and picking up bodies and carefully placing them in body bags. They were filling up shipping containers with as many as they could, and the containers would then be picked up by helicopter and taken to the ACT, where they would identify the bodies if possible. Otherwise, they cataloged them with a picture, DNA sample, and any belongings found on the body. They were keeping records so people could search for their missing loved ones."

The people just off the boat stared at him with looks of horror and disbelief.

He continued. "The news had it on every night and showed them picking up muddy bodies and counted how many containers they filled. Then the Melbourne tsunami happened, and at the same time, they showed the volcanoes in Indonesia were sending hot lava through the atmosphere, and a lot of people died there, too. What used to be a series of mountaintop islands had burnt to become a small black stub sticking out of the sea. They said Melbourne washed away and what land had been left at the top end of the Northern Territory burned and Australia is shrinking."

He paused, but nobody spoke, so he continued. "Then they showed a map of Australia the way it was fifty years ago and superimposed fire on the top end and where the tsunamis and rising water levels wiped out Sydney and Melbourne. It looks like Australia has been decimated and turned into a series of islands instead of one large mainland. Now they're looking for bodies washed out in the Melbourne tsunami. At least, that's what they say. It's not on the news as much as before. After awhile, they were only picking up and counting bodies every night, and it became boring news. The shoreline there now is the Snowy or Blue Mountains."

"Where are we now?" asked Ron.

"About ten kilometers outside of what used to be Wollongong."

"How long have you been here?"

"I've been living here ten years," Dan replied, "but the house is Ray and Tommy's. They're brothers and have been here their whole life and are in their fifties now. When it was still attached to the mainland, they had a lot of barbeques, and some of us living here now met back then. Their mom passed a few years ago. She had raised the brothers here on her own, and they both stayed and looked after her."

"Looks like a nice house," said Ron."I used to live in a similar house in Newcastle."

"Yeah, it is nice. The house was built by convicts in the early days of settlement." He pointed. "It's probably the only house remaining between here and the ACT over there, the Blue Mountains that way, and the Wattagan Mountains that way."

He told them that several years ago, they could see the mainland from the house, but the island their house sits on had actually slowly drifted out to sea, and with the land degradation over the years, the view of land was totally gone. He said that for a long time, it had seemed further and further away, and it was completely out of sight after the Sydney tsunami. He said the house was on a piece of solid rock that was on top of a long shelf. The shelf had shifted with the earthquakes and split, and they actually had moved more than ten nautical miles out to sea from their original location.

He said there had been a few islands with houses like theirs, but all were lost to the sea, and they'd only remained because they were atop a solid rock on a large shelf. He told them how after the Sydney tsunami, there was the Melbourne tsunami, and then the Tasmanian tsunami had hit at the same time the floodwater from the Americas and water from melting ice on Antarctica had raised the sea level again. He said Tasmania was now hundreds of small islands and only survivalists were there living on the high ground.

The people who lived in the house were survivalists of sorts. They grew their own fruit and vegetables and fished. They had a small sailboat they used to sometimes reach land or another island and take something to trade, but that had been before the tsunami, and they had stayed put since then. They had tank water and a wood-burning stove. They'd added the wooden shutters after a big wave had hit the house and broken the upstairs windows. Dan told them there were forty people living at the house. The brothers and their wives and kids were ten people who lived in the two bedrooms upstairs. Twenty people lived downstairs, and there were ten people who slept on the front porch. Most of them had been there for a very long time, and only a few had arrived recently after some of the nearby islands had been lost.

He said that it was crowded when they all wanted to sleep at the same time. They had trees and a vegetable garden in the backyard, and some nights, they could sleep out there under the stars, but it was cold in winter and no fun when it rained.

"Can we have a look inside?" asked Ron.

"No," said Dan."We can feed you and give you some water, but then you have to be on your way."

Ron recalled feeling the pain in his back and stomach and how he had been hoping to flatten out on a stationary bed with some kind of pillow. He was surprised at the answer after thinking they were all getting along so well. They had been out at sea for so long they were suffering, and he thought that should be a good reason for these people to let them stay. He was wondering why they would stop them from coming into the house, but after the rejection, he didn't really care to go in. He was still hoping to put his feet on the ground and for a flat place to sleep, and they could manage that outside. He thought about arguing but thought better of it when he considered they were going to give them food and water. He didn't say a word as he stood there looking at Dan while trying to ignore the pain returning to his back.

Dan continued."We've had some bad luck with visitors before. More so years ago, when there were a few other islands and parts of Wollongong and Sydney had more of a population and we would see more of them. Some kids would get together and act like pirates, going around to places like ours in little tinnies and robbing people. Every so often, we had to deal with them here. There have been other desperate types trying to rob us, too, and that's why we don't let people inside."

Ron looked at him and said, "We've been lost at sea and are just trying to reach land."

"Some people have been here for a visit and caused trouble or not wanted to leave, and they had to make rules. Ray and Tommy decide who stays and for how long."

There was a silent pause before he continued talking. He explained they were already too crowded and it was impossible to accommodate more. He said there had been people living there who had drowned or who had been taken by sharks, and when their numbers went down, they had more room and took others in. He said, usually, it was one of their friends or family who was living rough who would join them, and their numbers would go up again, but at that moment, they were at their limit.

"When was the last time anyone new came here to live?"asked Ron.

"Not long."

"How long since anyone died here?"

"Last year. Henry was the last one, and he was taken by a shark."

"Do a lot of you get taken by sharks?"

"The sharks have been pretty prolific the past four or five years and have taken a lot of people around here, though not so many that lived in this house. They've taken six people that lived here in the last ten years but probably a hundred others that were just visiting or nearby when there were other islands around."

A woman with a platter full of cooked prawns came out of the house and said, "You all can come up here on the porch while we eat."

"Let's get off these steps," said Ron, "before the tide changes and the sharks get us."

He looked at the woman with the prawns. She looked to be mature and almost as old as he was. She was wearing blue denim shorts and a white top that buttoned at the front and exposed her bra. He liked her face, her body, and her smile. He thought there were so few women like her, his age and attractive in short shorts. He wondered if she were single and if he should propose marriage if she were. She offered Ron a prawn, which he took and thanked her.

"I'm Ron."

"I'm Pam."

"How long have you been living here?"

"Fifteen years."

Ron thought it doubtful that she was single with that answer, and he told her it looked like a nice place to live as he peeled his prawn and ate it. Dan leaned over and whispered to Ron that she was Ray's missus.

Mary had been talking to one of the other ladies about what she had eaten, or not eaten, in the past five weeks. The other lady went into the house and returned with a tray holding a bowl of cherry tomatoes, a bowl of sliced carrots, and a pile of lettuce leaves. They passed everyone a cherry tomato, a piece of carrot, and leaf of lettuce. Everyone made sounds of approval as they bit into their cherry tomatoes and carrots.

The people who were living in the house and those who had just gotten off the boat talked about their recent ordeals and how they had survived the tsunamis. The people who lived in the house told how they had survived for so long in a house on a rock in the middle of the ocean. They had been isolated since the sea levels had first risen, but until recently, they'd had neighbors on other islands within sight. They had looked out for each other to a certain degree. They would hop around the islands in a boat, fishing and trading. Earthquakes and the moving shelf had broken up some of the islands over the years, but up until the last tsunami, they had enjoyed having neighbors. Some of the people from the other islands were there living on the front porch, and others had made it to the mainland, but most were presumed dead.

They said they had heard the authorities had stopped recovering bodies around Melbourne because there were too few volunteers to process them and they didn't want to keep adding to a huge pile of corpses. They said the authorities kept stuff like that out of the news.

Ron said, "How do you get information that's not on the news?"

"We have a two-way radio," said Dan. "Tommy and Ray have been on it since they were kids and have friends in Canberra that have them, too."

Pam was standing within earshot, and she butted in. "Ray already radioed emergency services and told them about you, and they said they are swamped and to hold tight until they can get back to us, probably a week at least."

"Tommy already said they should go when they're done eating," said Dan.

"Would it be ok if we stayed on land overnight and left in the morning?" asked Ron. "We could sleep on the rocks or around the house in the yard."

"Others tried to live on the rocks and were washed out to sea or taken by sharks," Dan replied. "When it gets rough or there's a lot of rain, they want to get in, and the house gets too crowded, and that's why Ray and Tommy have the say on how many and who stays or goes."

Ron nodded in acceptance, and Pam said, "Ray said he likes you guys, though, so I'll go ask him."

Ron joined the others and heard them tell stories about things their group had done and how long they had lived in the house. They were talking about huge hauls of fish and how they used to trade it for all kinds of things. They set crab and crayfish traps all around the island and would dive in and retrieve them in the mornings. They had to keep an eye out for the sharks and had stories of close calls and others who were taken. They said the sharks hadn't been as bad before the seas had risen, but since then, they had become progressively worse. There were more sharks now, and they were more aggressive.

Ron mentioned that when they'd arrived, he'd seen them in the water with the prawn net. One of them answered and said they had been keeping an eye out and were being quick about it. He said the sharks that were following their boat on the way in would probably be hanging around still waiting for something to move but would eventually go away as long as they stayed away from the edge. Ron thought that was another reason to hope they would be allowed to stay the night.

Ron looked to the upstairs window when he saw something move. A man and two children looked out and then moved away. He thought that was probably who was making the decision if they could stay the night or had to go. He wondered what would happen if he just said he was staying and that was the end of it. How did this character upstairs know he didn't have a gun? What if he walked upstairs with a gun and said they were staying? He could throw this Tommy and Ray out and stay there and make the house his own. He didn't really want to stay any longer than overnight, but being forced back out to sea seemed unreasonable. If the situation were reversed, he would let them stay. Then he thought about what if there were hundreds or more who had wanted to stay and about some of the stories they had told about how others had tried to take over or rob them. Then he thought they probably had a gun in the house. They had managed to stay in the house all this time and survive, so they wouldn't be pushovers. If they didn't have a gun, they would have weapons of some sort and be able to back it up when they told someone to leave.

Then Ron had a thought that Ray and Tommy were thinking the same thing about whether he had a gun or what they would have to do if there was trouble. In Ron's mind, it was a short telepathic conversation. He thought he didn't believe in such things, but the thought was there, and the reasonable action he should take was clear because of it. He would be grateful for their hospitality up to that point, and if they said they needed to get back in the boat and leave right away, he would thank them and shove off. There was no point doing anything else. Even if he wanted to overpower them, there were more of them, and it was a dumb idea. He was disappointed with their attitude and thought if the situation was reversed, he would be more humane, but he thought it better to keep his thoughts to himself.

Pam came out of the house and said that because Tommy and Ray liked the people on the boat and were impressed by how they had survived for so long out at sea, they could stay until morning. She said they should sleep on the steps and the jetty. She advised them to use large rocks and place them on the edge of the jetty and before the steps. That way, if the sharks rolled up, they would feel the rocks and wouldn't be going up and over them. She advised them to move slowly and stay quiet and all would probably be well. She said they were going to fill their water container and give each of them an orange when they left.

She said Ray had radioed again and had informed emergency services there would be a boat with thirteen people heading to the ACT and should be on land before dark tomorrow. She said Ray and Tommy had been watching and listening from the window upstairs and liked what they had heard, so they'd been happy to radio emergency services to help.

She said another thing they did there was winemaking, and they would open a few bottles, and everyone could have a glass and make it a little party. That started the conversations rolling again, and the people living in the house were happy to have company and a rare declaration of a party. It turned out they knew of the rubbish pile and would take a boat there every so often and look for material to bring back. They had found timber and bits of steel. They said it had been there since the sea levels had started to rise and had grown bigger every year. It had started out with a cruise ship running on top of two container ships and sinking all three of them. There was a current that ran over the top of them, and other ships that had gone adrift would run into those and sink there. Some containers that were watertight floated to the surface, and all kinds of other boats were sunk there. They had opened containers and found useful treasure from time to time.

The rubbish floating in the sea had started to collect, and all the oil and sludge had bonded it to the pile. They said the middle of the pile was covered with sludge. There had been some nuclear waste that had leaked out or something else that was killing everything on the south side of the pile. They said they had seen nuclear fallout in the atmosphere, and it looked like heat waves you could actually see. When it drifted over a flying bird, the bird would make a few slow flaps and drop dead. There had been a huge fish kill, and ever since then, they were afraid to eat the fish there. Ron mentioned they'd noticed a dead bird and that there had been no fish biting, and it was lucky they hadn't caught something there and poisoned themselves.

Dan said, "We noticed you folks have some timber in there. You should hit land tomorrow night and shouldn't need it." "How about we trade you some crabs for it?"

Ron thought they might be a little hasty and about what might happen if they didn't find land the next day. The discussion went around in relay, and in the end, they kept enough wood for one fire and traded the rest for sixteen cooked crabs. They ate the crabs, and with the house's homemade wine, which they shared for free, they all had a feast on the front porch. People from inside the house had come outside, and they passed around three or four bottles of wine. There were still others who stayed in the house and two who played acoustic guitars. The front porch and front steps were packed. Ron was elated and reflected on where they had been and how good he felt at that moment. It was a party he was really enjoying, and as he looked around, he could see everyone else was, too.

They mingled and talked until late and then started to prepare for the night. Everyone who had been in the rowboat for five weeks had stomach pains from eating so much and was tired, but they were relieved to know they would be on land the next day. Pam said it was lucky they had such a clear, warm night. She passed out oranges for everyone and suggested they eat them in the morning before they shoved off sometime before sunrise. Ron still didn't like the idea they were being forced to keep moving and so early. He thought the people were kind to provide feed and water and let them stay the night and to radio emergency services. He wondered why they would then be unkind and make them go before sunrise, and he wondered what would happen if they slept in. His inner voice told him not to test them. The two guys living upstairs had managed to hang on to the house when others had tried to take it from them. The people in the house were only being careful, and so they should be according to some of the stories they told. Ron pictured himself sleeping and then being woken while being attacked with an axe and the assailant screaming for them to get out because the sun was up.

Ron said to Pam, "Ok, give us a hug, and we'll be out before sunrise. Thanks for feeding us and letting us sleep the night."

Pam hugged him and said, "That's okay. You guys get to Canberra safe."

The hugs and wishes for good luck went around all the people on the front porch and the thirteen from the boat. The people on the porch closed a little gate at the top of the steps, and some started rolling out sleeping bags as others went inside. Some of them stood or sat around talking quietly.

Ron crept silently into the boat, and being quiet so he wouldn't attract sharks, he passed out the sleeping bags and blankets to those who had one. He passed the bits of Styrofoam or clothing they had been using for pillows, and they all found a spot on the pier inside the rocks they had spread around the edges. They slept four wide, Jin and Julie with the kids, then Steve and Deb with Charlie and Sue, and Mark and Amy with Bob and Mary, in three rows that took up the entire length and width of the pier. Ron wanted to sleep on a stationary bed of some sort, but the steps weren't wide enough for him to flatten his back, so he settled for the tarp inside the boat. He unrolled it and then folded it so it would fit on the floor of the boat and he could stretch out. It would still rock around and bump into the pier, but he could stretch out, and that was great. He felt severe pain in his back and heard a cracking noise as he flattened out, and then his stomach cramped and ached. He put a shirt under his neck for a pillow and opened up a jacket to use as a blanket. He was a little more vulnerable to the sharks than the rest of them, but he thought that the sharks had left the scene and that if he stayed quiet on the floor and under the jacket, he should go unnoticed until the morning. A few minutes later, he was sleeping soundly.

The next morning, Ron opened his eyes and was looking at the sky before he remembered where they were. It was light, and when he recalled they had been told to leave before sunrise, he sprang into action. He had started to roll over when the pain paralyzed him. He was in such a hurry he had forgotten all about his lousy back. He took slow breaths to relax, and a few minutes later, he carefully rolled over. It wasn't bright yet, so maybe, technically, the sun hadn't risen. He took a few deep breaths before he rose to his knees, and he was careful to keep his back straight when he stood up. He quietly climbed out of the boat and onto the jetty and started waking the others.

There were two people on the front porch who were awake and drinking warm cups of something. They were looking out at sea and at the people starting to wake and get off the pier and into the boat. The people on the front porch waved and gave the "OK" sign and called out that they didn't see any sharks. Everyone who was leaving relieved themselves on the rocks before they climbed into the boat.

As soon as they were in the boat, Ron untied the rope at the back, and Steve untied the knot at the front, and they pushed away from the pier. There were a few others who had woken and were on the porch waving, and two young children were waving from the upstairs window. Ron and Bob took the oars and started out slowly before finding a steady rhythm. They had been given advice and directions, so they knew the way to line up the house to stay on a steady course for as long as they could see it, and when they couldn't see it, they would probably see land in the other direction, or if not, they would follow the sunset and see land soon.

Ron and Bob paddled for four hours, and they could still see the house but not land. They stopped, and everyone ate their orange. Mark and Steve took the oars while the rest of them settled to keep their eyes on the house and make sure they were going in the right direction. They sighted the tree in the backyard, at the corner of the house, and they stayed steadily on course until it was a dot and eventually disappeared.

It was cloudy, but they felt it was near midday. The swell was growing, and the boat was going up and down. When they were at the top of a wave, they all looked for the shore. It looked like smoke in the distance, but they couldn't see land. They changed seats and took turns with the oars all day. They had spotted rubbish floating and several shipping containers that must have been airtight to be floating just beneath the surface. The plastic wrappers floating on top of the water looked to have brighter colors than the wrappers they had seen in the rubbish island, so they guessed the litter where they were had been in the water for less time and they were getting closer to the mainland.

They thought probably all the litter and debris they saw was floating out from the mainland and would eventually drift to Rubbish Island. They joked that there would eventually be a pile of rubbish large enough for them to walk across the ocean to the rubbish island where they had been. There were also shipping containers floating on top or just below the surface. They speculated that the shipping containers would be fun to open to see what was inside. The people the night before had told them how they would dive underwater to get inside a few of them and had once found one full of canned food. That had been a few years prior, and it had fed many of them for a long time. They had traded cases of canned food like currency, and unlike other barter items at that time, everyone was happy to trade for canned food. They joked about towing one of the shipping containers in with them, but all agreed they didn't want to be slowed down. None of them wanted to spend another night out at sea unless they had to.

When the dark arrived, the boat was rolling up and down. They stopped it and looked where they expected the mainland to be.

"I think I see light," said Ron."Do you guys think there's some light in the sky over there?"

"Yes," said Steve, and soon, everyone concurred.

They kept moving toward the light and took turns with the oars. When they stopped to swap seats, they could feel the current taking the boat in the opposite direction. They decided they would paddle all night, and if the progress was slow, at least they wouldn't be going any further out to sea. They kept moving, and as the hours passed, they began to see what looked like flames in the darkness. None of them were certain what they were seeing, but they speculated that it was a fire, and they guessed at what might be burning. They knew for sure it was at least giving them a beacon to aim for.

It started to rain, but they pressed on. Then it started to pour, and they couldn't see ahead, and the bottom of the boat was filling with water. They bailed out the bottom as they covered the boat with the tarp and waited for the rain to stop. It poured for hours. When it stopped raining, they uncovered the boat. It was still dark, and there was no sign of the sun or land. They proceeded in the direction they all agreed even though they were guessing it was the right direction. They could see sharks in the water that approached the boat every ten or twenty minutes, and they spotted several birds on top of the water. They surmised the more sharks and birds they saw, the closer to the mainland they must be and kept the boat moving as best as they could in complete darkness.

It was absolute darkness when the boat pitched to the top of a wave, and they looked ahead to see definite flames in the distance. They all made sounds that weren't put into words but indicated they had all seen the same thing. The next time the boat pitched over a wave, they saw it again, and they repeated their groans. The flames were a long way off, and nothing else was visible, but they were feeling excited and relieved. There was new energy and a call to concentrate on the job at hand, which was to make progress without being capsized. When they changed hands on the oars, they did it fast and kept the boat moving at all times throughout the night. When there was enough light, they could see smoke coming off the fire. When the sun was higher, they couldn't see the fire anymore but could see the smoke.

The sunrise was behind them, and they were heading toward the smoke. That started the jokes about going to the big smoke, and they sang "New York, New York," "On Top of Old Smokey," and "Light My Fire" and added their own humor to as many of the verses as they could. They were elated and ignored the sharks that would come near them from time to time. They kept their mind on the job and their eyes on the next wave between them and their destination. It was a jovial atmosphere and, at the same time, an intense effort to keep the boat from tipping them out while moving toward the mainland at all times. After all the time out at sea and being so close to safety, it would be horrible to be tipped out of the boat at the last minute.

When it clouded over and started raining again, they just kept moving. They didn't need the sun for direction as long as they could see the fire. They were soaking wet, and whoever wasn't rowing or keeping an eye on the front of the boat was bailing out rainwater

It poured and let up a few times for two or three hours before the sun reappeared. They guessed it was near noon, and they looked at the blue sky behind them and the smoke in front of them. The smoke made a huge brown haze on the horizon, and then they could smell it. There were different smells, some of them pleasant, like a burning campfire, and a few minutes later, they would get a whiff of something similar to rubber burning. There was more and more rubbish floating around, and it was becoming a task to miss some of the larger things floating in the water. There was the odd shipping container or capsized boat they had to miss while riding the swell and trying to keep the boat from tipping over.

The sharks had been constant companions since they'd first appeared, but they hadn't attacked or caused them any concern. When a huge white pointer opened its mouth and had another go at the back of the boat, it caught everyone by surprise. The top of its mouth chomped down on the hardwood, and it broke its teeth before coming partly snagged by its other teeth and shaking its head before dropping back into the water. It wasn't the first time a shark had put its mouth on the boat, but this one left deeper gouges than the previous one. They were surprised to see four other sharks heading for the boat and relieved when they stopped. It seemed the other sharks lost interest when the shark that had taken a bite apparently didn't find it tasty.

The rubbish they were paddling through was getting thicker, and soon, it covered the water. The boat would push apart plastic bottles and floating rubbish and show blue water underneath. The oars pushed back the water and the floating rubbish, too. When they went over the crest of a wave, they saw a new landscape ahead. There were endless mudflats after the surf breaks. There were shapes covered in mud sticking out here and there, but all they could see in front of them was mud and, beyond that, smoke. They paddled the boat until they could surf a wave to the edge of the shore. The wave pushed the boat onto the mud, and they kept the oars moving for several strokes before they were stuck fast.

Ron took an oar and pushed it straight down into the mud. He pushed it deep until only the handle was above the surface. He declared it was too deep to walk through, and he struggled to pull the oar back out of the mud, as it seemed to have suction holding it back. They discussed their options, which were few. They decided they would wait for someone to spot them. If they did try to walk out, it would be dark before they could see anything, and it seemed a better option to sleep the night and decide the next move in the morning. It might be lighter, and then perhaps they could see something. Also, because they were so much closer to land, they had hope someone might spot and rescue them.

When one of them had to defecate over the edge of the boat and the feces remained where it had landed, they used an oar to move the boat slightly. It wasn't like rowing the boat in the traditional manner, but they realized it was possible to push the boat about a meter and then lift the oars out of the mud and put them back in and move another meter. It was slow going, but they traveled a few hundred meters before dark. They looked back to the ocean and could see waves full of rubbish. They looked ahead, and they saw a vast mudflat with some smoke in the distance, and the darkness allowed them to see what appeared to be flames.

Ron thought that at least on top of the mud, the boat would remain still and make it easier to sleep. They all found their spots to bed down and made small talk as they watched the sunset. They were encouraged when they saw flocks of birds fly over them, and all agreed they would be on dry land soon. They knew they might have some more struggles and couldn't rely on being rescued, but they were certain they would survive. They reminded themselves what they had been through and congratulated each other for surviving.

Ron said, "I'm going to try and sleep. It's my notion that when you have a problem you can't do anything about until the morning, you may as well stop worrying and get some sleep."

"Yeah, said Bob, I think we heard you mention that before," and there was light laughter from a few of the others.

Ron flattened his back, rested the inside of his knees on the side of the boat, and let his feet hang over the side. They didn't have to worry about sharks or falling overboard, so it was ok to leave his feet out there. If it started raining and they wanted to cover the boat, he would have to bring his feet inside and sleep with his knees up. He would probably have to change his position after a while, but at that moment, he was comfortable. It was a warm, calm night with just a light breeze and the smell of smoke in the air, and they were all feeling relieved.

He thought how lucky they had been and how much he was going to enjoy sleeping without the boat rocking and waking him up every few minutes. He closed his eyes and let his mind ramble to forget about his back pain and tried to relax his muscles. He could still feel the sensations shooting up and down his back, and then they nearly stopped before he fell asleep a few minutes later.

The next morning, Ron woke up at the crack of dawn. The first thing he thought of was his surprise for sleeping so soundly without interruption, and the next thing was it might not have been a good idea to sleep with his feet over the side of the boat. His feet were numb, and from the back of his head to his neck and shoulders and under his arms was also numb. His fingertips felt like pins and needles. He struggled to lift his feet and straighten his arms and legs. He moved his head and stretched his neck as he shook his hands and fingers, trying to loosen them up and get some feeling back. When he turned over onto his stomach, his face and head felt like pins and needles, and he feared he was passing out. He remained paralyzed for a few minutes before the feeling passed, and he stood up. The boat in the mud had remained steady all night, but when he stood to one side, it dipped, and he had to step back and get his balance. That woke a few of the others, who opened their eyes and began to move about.

Bob called out, "What's going on?"

"We're still here," said Ron.

There was light laughter at first, and when Bob retold it when Charlie and Sue woke, and then again when Mark and Amy woke, it cracked everyone up. It was still too dark to see much, but it was so calm and quiet they could hear the surf behind them and smell the smoke in front of them, so they decided to take off anyway. It would be slow going in the mud, but they wanted to keep moving and were hoping their ordeal would end soon.

Chapter Seven
Don't Be a Stick in the Mud

The oars couldn't be used in the usual manner to get through the mud, but they made progress. The mud became thicker as they moved away from the ocean, and their travel became progressively slower. They found the best spot for someone to stand was with an oar on each side of the boat at the middle of the front half. They would hold the oar straight up and down over the side and then stab it into the mud as deep as the paddle and use the mud to push off and move inches, or maybe a foot, at a time. It was slow going because, as they brought the oar back out of the mud, it collected mud that they had to scrape off before putting it back in to go just a little further each time. They found a knack of scraping the mud off on top of the other mud as they lifted them out and began a rhythm.

It was slow going, but they kept moving and started a chant of "Go!" whenever someone was about to push forward with the oars. As soon as anyone looked like they couldn't keep up with the rhythm, someone took their place. The sun came up and revealed more mud for as far as they could see. They looked back at the sea to how far they had come since they'd woken and guessed it was another hundred meters. They looked ahead and guessed it was at least ten times further to the edge of the fire. They weren't sure what they would do when they reached the fire, but they considered the immediate task was to keep going. They kept moving at their slow and steady pace for another four hours. They looked back to see how far they had come and were happy with their effort. It had been hard work and slow going, but it was definite progress in the right direction. They could see off in the distance piles of debris, with cars, buses, and something or other sticking out of the mud like hundreds of little ant hills. They could see smoke, but they could also see trees on a hill beyond that.

They were taking turns working and resting. Ron worked hard, and he fell asleep when it was his turn to rest. He woke suddenly and in a bit of panic. His pain seemed to be gone ,at least for the moment, but he couldn't move. His eyes were open, and he was trying to get up, but the feeling he'd had before like he was paralyzed or magnetized to the ground returned. He felt numb and tried to yell but couldn't make a sound. He closed his eyes and tried to relax and fell back to sleep.

When he opened his eyes, he didn't know how long he had been sleeping. He was relieved when he next woke and could move and sit up on the bench.

They were taking a break and having a cup of water when they saw something in the sky. It was a small drone, and when it was close enough, they could see it was a police drone. They started cheering and waving, and it came straight for them. It stopped in front of them ,and the light started flashing and going around in a circle. There was a screen with a rolling text that read: "Help is on its way. Approx. two hours for the next available rescue helicopter to reach you."

Then the drone sped off as they started hugging, and some of them were crying tears of joy.

There were calls of "We made it!" "I knew we would!" and "Yahoo!"

They all drank another cup of water, and Ron tried to start a sing-a-long. They all enjoyed a verse of "Roll out the Barrels," but they were too excited to join in anything else. Ron sang a song he made up at that moment: "I'm going home to see my baby doll." He started a rhythm with his hand on the side of the boat, and some of the others joined in. They were chanting and clapping and entertaining each other as they waited for the chopper.

The children did some dancing, and then the girls joined in, and soon, everyone was standing up in the boat, doing the twist and clapping their hands. They all agreed they would stay in touch and get together for a shindig once in awhile once they were all settled back on the mainland.

They were enjoying the moment when they heard the chopper in the background. They all waved at it as it approached. It hovered over them, and two search and rescue police dressed in white overalls with blue helmets dropped down into the boat. They introduced themselves and asked if anyone was hurt. Sue said that Charlie had been the only one and that he was a lot better but still suffering from the bruising. The police had jackets with a harness around them, and they put one on Sue and Charlie. There was a loop at the back they used to attach a hook, and they were winched into the helicopter. The empty jackets on the end of the hook returned immediately. They were lifted out two at a time, and when all the survivors were in the chopper, they sent the hook down for the police who had put them all into the jackets and were waiting in their boat.

When the chopper lifted high into the air, Ron looked out, and he saw a view that explained a lot to him about what had happened while they'd been out at sea. The fire was oil that had drained and settled at the edge of the mudflats after the tsunami. Volcanoes at the top end had sparked a huge bushfire that had island hopped and covered the mountains from Dorrigo to Penrith. The tsunami had covered everything to the foot of the mountains. The trees on the mountains were on fire and spreading burning ash. The oil had ignited in various places on the edge of the mudflat. It was a burning barrier at least a hundred meters wide between the mudflat and burned-out land. The fire stretched from the edge of the Barrington Mountains to Penrith. There was nothing that could be done, so the authorities had declared they would wait for the fire to burn itself out and then consider a clean-up.

Ron looked up and down the coast for anything he could recognize. He could see part of the Sydney Harbour Bridge and Centre-point Tower sticking out of the mud. He could see other buildings, too, but it was only the bridge and the tower that told him that he was looking at Sydney, or what used to be Sydney. He could see the fire was in the trees in the distance and on the surface of the mud where it met the shore. He could see green trees on the other side of the smoke and mountains beyond that. Then they were above the clouds, and he couldn't see the ground. He lay down on the floor of the chopper. He could straighten out his legs, and his back was flat. His back was hurting, but overall, he felt good and cracked a huge smile as he drifted off to sleep.

Approximately one hour later, the chopper was landing. The change in the air pressure, or perhaps the sound and movement, all worked to startle them from their sleep.

"What did I miss?" asked Ron.

"We're at the Canberra Airport," said Bob. "There are cars and houses and people about. It looks like life as usual here."

The policeman sitting next to him spoke up and told how they'd had severe storms and floods that had done a lot of damage, but the clean-up had started the next day, and everything had been open and back to normal in less than a week. Then another cop stood up as the chopper set down and the engine started to quieten. He said it had been pretty hectic and the hospitals were full. He said a bus would pick them up and take them to a school auditorium that had been converted into an emergency center, and doctors would look at them there. He said the ACT had closed its borders and that people trying to reach the

ACT had to show they were residents or staying with residents or they were turned back.

Ron said, "I thought there was a citizens group claiming all Australian citizens and residents were allowed to camp on the grounds of Parliament House and they started a camp."

"They did," replied the policeman, "and there are a million or more all cramped in and living on top of each other. They're in tents on the lawn and in the car parks."

The police checked vehicles on the road before they reached the driveway, and they acted as doormen at the entrances to Parliament House. A lot of police and politicians and their families were living inside Parliament House. Those that had duties there could live on the premises and were given an office they would convert into a living area. Parliament House was considered the safest place in the country, so many of the workers had moved their wives and children inside. The families had one room to live in and shared a common kitchen, dining area, toilet, and showers with the other employees' families.

Ron told the cop he could stay with his son and family at their house. Bob and Mary spoke up and said they had someone to stay with, as did everyone else who had been in the boat. The cop said that was good because, otherwise, they didn't know what they were going to do with them all after they left the emergency center.

The rotor slowed and then stopped completely before the cops opened the doors. They carried Charlie out on a stretcher, and as soon he was a few feet away from the chopper, several people, some with microphones and cameras, converged on him. The cops helped the rest of them climb out of the chopper one at a time. The children and Jin and Julie went next, and then the rest filed out two at a time, and Ron was the last one out.

The media were there and asking the others questions as he walked with them. The police had shown the media the video recorded by the drone of them struggling a few inches at a time to get across the mud. That was all they had seen, and they had concluded they had been in the mud the entire time since the tsunami had hit Sydney. Steve was trying to tell them what had actually happened, and he managed to get through to them just before the cops asked everyone to get on the minibus because there were other emergencies they needed to attend, too. The media rushed to their vehicle to follow the bus so they could continue the interviews at the emergency center. As soon as everyone sat down in the bus, it started moving.

"How about that," said Steve. "They say we're an epic story of survival and are going to be on the news. They had given up the search for survivors weeks ago and were only looking for bodies."

Chapter Eight
All's Well and a Story to Tell

The bus pulled into the emergency center, which had been a school, and they drove around to the back of the building, where it stopped next to two closed doors. As soon as the police stepped off the bus, the school's doors opened, and four teenage kids came running out. One of them had a stretcher they used to carry Charlie off the bus. The rest of the survivors walked off the bus and made their way inside to a basketball court with folding tables and people all around. There were people in beds with drips on the court's floor and people sitting in groups in the bleachers. Half of the basketball court was filled with two dozen stretcher beds with patients, nurses, and doctors. The other half of the court had folding tables and chairs and a line that everyone off the boat was now at the end of.

Charlie went in on the stretcher, and they moved him straight to the front of the line. The doctor had a look at him and made a few motions with his finger to indicate to a ward to take this one over there. The ward pushed the stretcher to a line with three others near half-court. Some of the media people who had met them at the airport walked in and started filming everything that was going on in the center. They found Steve and asked him to continue his story, and he obliged.

Ron was waiting in line when he noticed a man with a clipboard talking to people in the line from the front and making his way to the back. He appeared to be asking them questions and writing down answers. He had a flashlight and looked into a man's ears and used a stethoscope to listen to a few people's hearts or lungs. A few times, he wrote something and handed it to the person, and they walked off, making the line a little shorter.

All the others were in line in front of Ron, and he heard a man talk to Sue and say he was a doctor. Sue said she was ok, and the doctor said she could go, but she said she wanted to stay with Charlie. The doctor talked to them all and gave them all the same advice on what kind of food to eat and what they might expect after fasting and eating mostly fish for so long. He gave them all a script for a hydrating vitamin drink and gave Ron a script for his back pain. He said they were all clear to go and pointed to a door and told them they could go to the office and use a phone to call someone to pick them up or a taxi. Sue said she would wait there for Charlie, and the rest of them walked over to the office and waited in line.

The media was taking pictures and asking questions as the survivors were allowed into the office to use the telephone. They had become a good news story, and the media needed some good news after the tsunamis and earthquakes had killed millions and destroyed Sydney and Melbourne and almost the entire east coast. Queensland and the Northern Territory were deemed uninhabitable, and only a small number of survivalist on high ground remained. The Blue and Snowy Mountains were the new metropolises, and there were camps all around Uluru. Perth had been swallowed by the sea, but there was enough high ground to declare Western Australia habitable. Tornados had become common in central Australia and destroyed a community from time to time. South Australia had been inundated after the rising sea levels, and only survivalists on high ground remained. The ACT had lost some areas to the rising sea levels but remained much the same, and Canberra was considered the most livable city in Australia.

Steve did most of the talking when the media asked questions about how the thirteen of them had survived. At first, the media seemed disappointed they hadn't been stuck in the mud all that time, but when they heard of the group's freakish near-fatal events and how they had been lost at sea, they became much more interested. Steve had the media intensely hanging on to every word he said when Ron took his turn to use the phone and called his son. He left a message and told him he was ok and at the school and would call a taxi and be waiting for him at his house when he got home from work.

The others all called whomever they knew in Canberra and found places they could stay. They were exchanging their new phone numbers and saying goodbye to the media when a man in a suit walked up to them. He introduced himself as Malcolm and said he worked for the federal minister for the environment. He asked which one of them was Ron, and Ron raised his hand. He said there was someone at Parliament House who wanted to talk to him. Ron said he had just called his son and was on his way to his house. Malcolm assured Ron he could call his son from Parliament House and that somebody there would give him a ride to his son's house when they were done talking. Ron wondered why someone would want to talk to him and how anyone from Parliament House even knew who he was.

"Who would want to talk to me?" he asked.

"His name is Bill Smit," Malcolm answered, "and he was a private sector executive and advisor to the minister."

"How does he know me?"

"I don't know. He saw some of the pictures from the media and said he wanted to talk to Ron."

"He knew my name?"

"Yes, he did, and he asked me to come over here and bring you to his office."

Ron hugged the others one by one, and they all said they would stay in touch before he and Malcolm walked off and made their way to the car in the parking lot. Ron sat in the passenger seat, looking out the window at the traffic and people walking or on bicycles. He said it was amazing that life looked so normal there when it was reported to be chaos everywhere else. Malcolm said they'd had floods but had been lucky in the ACT and in Australia in general. He said pretty much everywhere else in the world, they were worse off. Civilization had been shrunk to the point most governments had folded up, and survivalists living on high ground off the bush outnumbered people living in cities.

When they reached Parliament House, Ron saw there were people and tents covering the lawn. The guards recognized Malcolm, and they opened the gate to wave him in. He drove under Parliament House and through a parking lot without cars where more people were camping. They didn't all have tents, but there were blankets on the ground, and some areas had tables and chairs. Malcolm drove past them and into another parking lot full of cars. Ron saw only one vacant parking space, and Malcolm drove straight to it and parked.

"Lucky there was a spot," said Ron.

"It's this car's spot," replied Malcolm. "It's reserved parking only, and each car has its own space."

He explained how the civil libertarians had tried to keep all the cars out of Parliament House, saying it was more important for people to have shelter. He said there were two hundred cars and seven drivers for every car. They each had their own day of the week to use the car for their business. He said it was the only car park at Parliament House that people weren't living in. He said the cost of registering a car in the ACT had become so great that the government had gotten rid of most of its fleet, and the cars they'd kept were seldom used anyway, and that was why the parking lot was full. Ron had seen a number of cars on the road, but he had noticed there were fewer than when he had been there many years before.

He said, "I noticed there wasn't much traffic. Is that because of the registration?"

"Sometimes, the registration cost as much as the car," said Malcolm, "so people have given them up."

"So, who was driving all the cars we saw on the way over here?"

"People with money; Bill must think you're important to send a car for you."

They got out of the car, and Ron followed Malcolm as they walked up a staircase. They walked up another flight of stairs and down a long hallway. Malcolm stopped at one of the doors and knocked, and a voice said, "Come in." He opened the door, and they walked in. There was a short, thin man sitting behind a desk, with round wire-rim glasses and short, thinning gray hair. He was wearing a suit, and he introduced himself as Bill Smit as he extended his hand. Ron had a feeling he had seen this guy before, and as he shook his hand, he tried to remember where or when.

"We've never met," said Bill.

Malcolm excused himself and left. The office was around three square meters, with a desk and several chairs, a couch against the wall, and wooden shelves full of books and a TV. The TV was on the news channel with the sound on mute, and Ron stared at the beautiful newswoman on the screen.

Bill sat down behind the desk and offered Ron the seat in front of the desk. "We called your son, and he will be joining us soon."

Ron nodded. It was like this guy knew what he was thinking.

"I don't always know exactly what you're thinking," said Bill, "but I do know you're wondering why I asked you to come here."

Ron nodded, and Bill said, "Now, please remain calm when I tell you I'm from another planet."

Ron laughed nervously. He looked at Bill with doubt and said, "You must be sorry you came here."He looked Bill in the eyes. He didn't doubt him absolutely, but he expected an explanation that was rational.

"The state of the planet is why I'm here," said Bill.

He took his time and explained how the volcanoes, floods, tornados, hurricanes, and earthquakes had changed this planets landscape forever. Bill said they had been watching from their planet but were powerless to help or warn people on this earth.

Ron's head was spinning, and his hands were shaking. "Watching us from where?"

"From our solar system, where we have been watching you since the ice age killed off your dinosaurs."

Ron opened his eyes wide, remained silent, and looked Bill in the eyes, hoping he would give some indication whether he was bonkers or sane.

Bill said he had arrived with a crew of twelve others and they were only the second crew to ever visit this planet. He said his direct ancestor, whom they referred to as the "great-grandfather" had been on the first crew and had arrived in 1761. The first crew from the other planet had been a group of six people, and only his great-grandfather had remained, while the others had returned. He said his great-grandfather was many generations removed, but they referred to him as their great-grandfather. Bill said his great-grandfather had been a young man at the time, and when Bill had been a young boy, he'd heard of his feats and even seen his images from the other planet.

He said his great-grandfather had rescued women from slavery, including a Native American woman from a brothel, and they had raised a family and had seven children.

"One of them was your ancestor," he said, "and so, as it turns out, we are related."

"I think I believe you," replied Ron.

"What do you want to know? Just look at me."

Ron looked into Bill's eyes, and it was like his inner voice was asking and answering all his questions. Sometimes, it sounded like Bill's voice inside his head, and Bill would nod and smile like he knew everything Ron was thinking. Bill's planet was much the same as Earth. They had different languages, but their natural world was very much the same. There were twelve other solar systems other than their own that they had visited, many more they had detected, and an infinite number possible. All the solar systems that supported life were very much the same in many ways, but there were differences. All of them had been formed over indeterminable time when an energy source as great as a sun attracted and emitted other gases. Over time, the gases formed an orbit around the sun, and any other gas, liquid or debris it attracted would collect where possible. That flux eventually formed the planets, creating a solar system, which then produced life.

All the solar systems they had visited had the same planets in orbit at the same distances from the sun. There was only ever life on the third planet from the sun. The life on that planet started beneath the surface from what was essentially a huge compost pile of various bacteria. As the planets and the moons orbited around the sun, the planets collected all manner of matter in space. It was organic matter that collected on the third planet from the sun. Over millions or billions of years, meteors with various chemical properties crashed into the earth. The sun radiating onto the spinning planet created moisture when the heat of the day turned to the cool of the night, and then it was evaporated when the sun came out the next day. That produced clouds with rain, and after an indeterminate amount of time, the surface of the third planet from the sun resembled a tar pit or, you might say, a pile of compost. Then bacteria had to grow and die over and over every day and night until it mutated and survived longer each time. It would grow and then die, and the remains created new bacteria. When the bacteria survived, it grew cells, and there was more and different growth.

Everything was still beneath the surface when cells evolved to the extent they could procreate. There was one cell, and it would die. If the one cell could split, there would be two cells, but they would both die. If the cell grew another cell and the two cells grew others, it would keep growing; it would be procreating, and once it was procreating, the cells could split, and a species would be created. Once the planet reached that cycle, all manner of life would grow and procreate into different species, and life would go berserk. Any and every kind of animal or plant life that came into existence tried to survive and procreate. Once the third planet was at that stage, all forms of life were created, and most died.

Humans were the species that grew brains and were the dominant creature in all solar systems they knew of. Humans started as a cell. Then the cell separated but remained attached. Then, like a mother and a father procreating a child, inside of the placenta, another cell was created. Humans crawled out of the tar knowing they had to eat and procreate to survive. It was the collective consciousness of all those humans who had tried and died and created the DNA that every human after them would also carry. That was the thing humans called a soul. It was the DNA of human ancestors who still lived inside of the living. On all solar systems, humans were the same, but only on the planet where Bill was born were they conscious of the fact that they all were part of one soul. Ron and Bill shared ancestors and their DNA, and that was why Bill could communicate with telepathy.

On Bill's planet, they referred to it as the original planet, and the other solar systems were numbered in order of their discovery. They didn't know if the original planet was created before the other solar systems or if it was just that their ice age was millions of years before those in the other solar systems. Of the other solar systems they had visited, three had yet to experience an ice age. There could be potential for humans to escape a dying planet and settle on a planet before the ice age. There, they were still overrun with dinosaurs, and it was not known if or when a meteor would strike them, or even if they would also have an ice age. There were humans currently living on one of the planets that had been surviving with the dinosaurs for so long they had them under control. They hadn't killed off all the dinosaurs but had thinned them out and could keep them away from their communities. The human population there was sparse and non-existent on some of its continents.

Humans from a dying planet, like this planet earth, if they could travel through space, could live there, but they would have to survive the prehistoric conditions. The life on all the planets third from the sun in the other solar systems was always alike but with subtle differences. The original planet had never had wars or all the crimes, lies, and politics of this planet and others. They were never racist or had religions, and they thought that could be another indication their solar system was the original. Also, they had seen other solar systems destroyed and the life extinguished, and they had seen other solar systems that were destroyed before they produced life.

As human life grewinside a solar system, the galaxy it was in would grow larger. When a human was born, a new star in the galaxy was also born. The new star would reflect the life of the human that created it. The star would change color and shape, and it would shine brighter or become dull depending on the human's experiences and nature. Each star reflected the entire life of the individual human and could be read from afar by humans on the original planet just as humans on this planet read history books. The galaxies told the story of life in the solar system and could also be read like a history book to tell the reader which humans procreated and their experiences. The size of the galaxy indicated the number of humans that had procreated. The shape and color of the stars indicated how humans were related and what life experiences they had or were having.

On the original planet, they now had the technology to view some other planets directly, but previously, they had relied only on the stars to learn about life in other solar systems. The technology on the original planet had recently allowed them to study other solar systems in more detail, and they were still developing new technology and asking and answering more questions. They hoped that one day, they would be able to travel to an infinite number of solar systems.

The human life on all planets started the same. As humans rose from the tar, those closest to the equator were darker, and those further away were lighter, and those at the top and bottom of the planet were also darker. Those emerging in daylight were lighter than those emerging at night. Humans were black and white and every shade between. Human groups on this planet also migrated up rivers during conflicts over territory and differences in opinions about water usage and rights. The theory was that the sun was hotter at the equator, and at times, the sun on the poles was out almost all day, and these variances caused variations in skin pigmentation.

It was when humans evolved to the point they could produce skin and hair that they separated from their cells and began to explore. What they believed on the original planet was that the early humans encountered others of different color and embraced. On the original planet, they saw the difference in skin color as sexually attractive, and dark and white people procreated with each other from the earliest opportunity. They never had a war or harmed each other. To survive on the original planet, they worked together instead of competing. They embraced the living spirit that belonged to them all and their ancestors. The ancestors' bodies lived and died, but their DNA still existed inside of them all, and it was the same one living soul they all shared from the beginning of time that they shared now.

The reason human brains were so large and those on this earth said only part of it was used was that people on this planet didn't understand that the DNA of their ancestors lived in their brains and was how they were all connected to the one soul. It was the one soul they called life that had been the same living thing since they'd crawled out of the tar, and it connected every human, past and present.

On most of the other planets that the people from the original planet visited that produced life, they found the humans there had rejected others of different color in the first instance. They hadn't explored like the humans on the original planet. They only procreated with others of their own color, and they remained in the same location for many more generations. They spoke different languages and created different religions. They weren't in touch with their common soul and missed the opportunity to receive advice from their ancestors. They lied to and cheated one another. That led to wars. On the original planet, humans couldn't lie because they were all in touch with the common spirit and knew the truth. Humans on the original planet had never murdered or robbed one another either.

All solar systems that supported life had similar landscapes on the planets. The third planet from the sun always had oceans, mountains, trees, plants, rivers, and lakes, and only the shapes of the landmasses differed. This particular planet was the most violent and the only living one to produce nuclear weapons. Only other solar systems that had died had produced nuclear energy and/or weapons. Other emerging solar systems were destroyed when those solar systems died. As the dead planets and moons were flung from their orbits, they went through space, crashing into planets of other solar systems. Some others, like this earth, took a hit and suffered an ice age. Other emerging solar systems were destroyed before they could produce life.

Other matter was soaring through space as solids or breaking up and was still out there. That debris still out there would eventually crash into something or get sucked into another solar system and could be the catalyst for life in another solar system. So, even though a solar system died, the debris it sent through space had the potential to contribute to the creation of another solar system and life. The people on the original planet were conflicted with the notion they should intervene and try to save this planet earth, much like people on this planet earth were sometimes conflicted when it came to the intervention of wild animals. They decided it was the right thing to do for all humans on all solar systems.

In one of the other solar systems that had been destroyed, the humans had made their earth almost totally uninhabitable from nuclear fallout. Then they tried to survive on the moon, and that led to nuclear strikes on the moon, which eventually caused a huge split in the surface and enough heat and wind current when it was closest to the sun to move it out of its orbit. When the moon went out of its orbit, it started a warp and moved the planet out of its orbit, which carried on to the other planets and the collapse of that solar system. That caused the end of that solar system, plus two others that were supporting human life and at least seven others that had the potential to create life. A solar system might have an infinite life, or it might be destroyed at any time. They didn't have any means to intervene but watched it all unfold. They decided from then on they would intervene when they could but only for the good of the human race in all solar systems, not just their own. If they could save a solar system from exploding and potentially killing off other solar systems, including their own, they would assume that responsibility.

The original planet might or might not be any older than this planet, but humans had evolved sooner and differently, and it was much healthier and in no danger of self-destruction. Humans lived differently there, but physically, they were much the same from head to toe. They were non-violent and had never invented guns or gunpowder. They used hemp to make cloth, paper, oil, and rope and had lived in hemp tents when they'd first become civilized. They'd discovered coal and oil and used it for energy and heat but had never invented the combustion engine. They had discovered solar power hundreds of years later and switched to it because it was less effort.

They had different family units, with the women and children staying inside the community, doing chores and schooling, while the men lived outside and worked. The communities where the women and children stayed were fenced to keep them safe from wild animals. The men used to hunt wild animals in the prehistoric days but now kept them culled and on nature reserves. Men today still worked outside the fences, but at night, they went inside to socialize with the women. They lived in houses usually made of stone and had large yards, and everyone grew as much of their own food as they could. They had community gardens, which was how they described a farm, in every town. They grew quantities of food in the community garden and traded surplus with other communities. They grew different types of hemp in every community. Some were for paper or rope, and others were for medicine or oils.

They were very casual when it came to sex but very careful when it came to procreation. On the original planet, they were more in tune with their common soul than the others they had seen. It might seem like they read each other's minds, but it was the DNA of the ancestors, the human brain and its waves, and the common life and soul they all shared that allowed them to communicate without speaking.

Ron thought, *How does that work?* Then he thought, *Just like it is working now.*

When a person asked a question to his inner self and found the answer, it was their ancestors talking to them. Not just their ancestors, but the common soul that had been the one living thing since life commenced. Each person only lived once. They died off, and the human race, being the common soul, lived on. If they procreated, their DNA lived in their offspring and all ensuing children. When one child emitted thoughts or visions, all children with the DNA of that ancestor could receive them. It was not really their ancestor anymore, but the common soul they all shared.

Ron thought, *So, you die, but your common soul lives on? When someone thinks of you, you know it, but it's not really you; it's the common soul.* He thought he was beginning to understand the notion.

"Yeah," said Bill, "you are; we want you to come back to the original planet with us."

Is that possible? thought Ron. *And what does that have to do with you saving our planet?*

"We have the technology," replied Bill, "to travel through space using the solar power from our sun to exceed the speed of light. An actual flying saucer with continuous spinning that winds a fiber around a series of reels that turn over and unwind when they are full, keeping the ship in perpetual motion. We charted a course through galaxies and from star to star to refuel the solar power cells along the way."

Ron was amazed, but he still had doubts and wondered if it was a trick, although he couldn't imagine how or why this guy could read his mind and was talking to him. He didn't say a word as he sat still with his eyes staring straight ahead.

"It's not a trick," said Bill. "Look at me."

Through Bill, Ron learned that his people traveled through space and could go to any star they could see. They could see everything in any galaxy by reflecting light from their planet to the stars in other galaxies, and they could find solar systems that might hold life. The solar systems were like oases in a desert. They could actually bounce the light and view real images of those distant planets from their backyards.

Ron heard his inner voice change and sound like Bill. It said, *The time it takes to receive images of other planets varies depending on how far they are. The distance to this planet earth was less than most others, and images received were only seven days prior.*

They had recently created new technology to hear the audio, but previously, it had taken fourteen days to reach them, so they would view an event and have to wait two weeks for the sound. Ordinary people all over the original planet might be sitting around a picnic table in their home garden and looking at an event on this planet and then trying to guess what it was all about. They had to wait for the sound to know who was right, and that provided great amusement.

Before that, they didn't have any audio and had to speculate what people on this planet were doing or read their stars. People all around the original planet found looking at and speculating about events of humans and life in other solar systems entertaining. They didn't have television, plays, or movies, but they watched real life unfold on other planets. They had a particular interest in Ron and Bill's great-grandfather and the ensuing relatives. Humans from the original planet had visited other planets, but only on this earth and two other planets did they procreate, so they were particularly interested in life here. It was deemed not appropriate to procreate on any of the other planets.

When they discovered this earth's customs, intellect, and morals, they decided that if they were to procreate, they had a responsibility to the child and couldn't abandon it to return home. The physical journey between the planets in the different solar systems was only twenty-eight days, but it could only be made when the stars in the different galaxies and suns in the different solar systems lined up, and that was only once every seven years. They thought seven years was too soon to leave the child behind, and it was not appropriate to take the child with them. There was one man who had volunteered to stay on this planet for the sake of humanity everywhere, and that was the man they referred to as the great-grandfather. He'd rescued women who were being abused, including the Indian woman who had been forced to work in a brothel. He'd taken her and other women as his wife, and they'd had children. All of their children and grand and great-grand children's offspring, like Ron, shared the DNA of a human from the original planet.

At that time, their aim of procreation had been to determine the similarities and differences of humans from different solar systems. The great-grandfather taught his children how to play the piano and how to make candy, among other things. One of his offspring made his living making and selling candy from his shop at the front of the family home. He could have done anything and become a leader or a celebrity or rich, but he chose to eke out a living, and he enjoyed raising children. His responsibility was to give the children a happy childhood and a fair start in life. He thought children loved candy and music was fun and made people happy. Two of his children, brothers, started their own candy store when they were older, and all of those children, including Ron's father, learned to play the piano and how to make candy. Ron recalled what he knew of his family history, and it all seemed possible.

Bill said the people on the original planet had been watching Ron and all of his living relatives. There were currently a few thousand living beings on this planet who shared the DNA of the people on the original planet. Every one of them had higher-than-average intelligence and enjoyed better health and the ability to heal quicker than the rest of the population. The original study had been to determine if human procreation between the planets was physically possible and if the offspring would be born with original knowledge of both planets. The term 'original knowledge' was what humans called 'animal instinct'. An infant in the womb was connected to the mother's brain, and even though it couldn't see, speak, or hear, it knew where it was and what the outside world it would be born into was. They wanted to know if the unborn child would recall the original earth, and incidentally, they discovered that the humans who shared that DNA were above-average specimens.

Bill landed on this planet, and as soon as he found his first relative and spoke to him, it was like a spiritual reunion with his great-grandfather. He understood this earth and its humans' history and evolution, and in his inner eye, he saw every relative and their inner voices and life history. The humans on this planet were not born with the human instinct of those on the original planet, but those who shared the DNA were more intuitive.

Ron thought, *Yeah, all my relatives seemed smart and intuitive, and it was like they had some kind of connection whenever they saw each other.*

Then he heard Bill's voice in his head. It said, *The people on the original planet would like to study if the same transformation would take place if a human from this planet reached the original planet and procreated with the population there. The bloodline of the great-grandfather is becoming rare on the original planet and the experiment would have the potential to ensure the bloodline and that life in general lives on. Also, the women on the original planet currently greatly outnumber the men.*

Ron's inner voice, still sounding like Bill, continued. *Thousands of humans on this planet earth currently sharing the DNA from the great-grandfather are infants to elderly. Only eighty-four of them are single, capable men older than forty. Of the men who share the DNA, only those who've had vasectomies are incapable of procreating. We want someone over forty years of age because the men under forty need to remain and rebuild society on this planet.*

Ron wondered in his own inner voice why they wanted him instead of one of the younger ones, and his own inner voice told him it was because older men created children with greater intellect. Ron was also one of the last remaining unmarried and capable men of his generation. Based on physical condition, state of mind, and weighing up all the other factors they could think of, Ron was the top pick. They had been watching him and the others since he was a child, and since the time Bill had been on the planet, he'd been able to hear Ron's inner voice.

People on the original planet appreciated Ron's outlook on life and how he continued to survive and keep a positive outlook in dire situations. They appreciated the way he played the piano and wanted him to come to their planet and play for them. They liked the way he was attracted to beautiful women and his respect for them. They liked his inner struggles when he was around an attractive married woman and how he never acted inappropriately with them. Every time there had been a choice between good or bad, constructive or destructive, Ron had made the right choice. In a world full of liars, cheaters, and robbers, Ron would not lie, cheat, or steal, and they liked that.

They considered it had a lot to do with the DNA. They took procreation seriously on the original planet and agreed Ron would be the best choice to procreate with humans from the original planet. There was a worldwide website on the original planet, and hundreds of thousands of women had signed up to procreate with him if he agreed to travel there. There were six women who had arrived with Bill and were returning to the original planet, and they would also like to procreate with Ron.

"Now this is all getting too bizarre," said Ron. "I must have drowned out at sea and am dreaming about all this from a watery grave."

"No."

"How will we travel to the original planet? In some kind of spinning top or flying saucer?"

Bill laughed and said, "The spinning reel is the outer shell, and the inner shell remains still." Ron looked puzzled, and Bill said, "I'm staying here and hope to procreate and raise a family. If you go, it will be for at least seven years. You may decide to stay and live out your life on the original planet, or you can return in seven years. It will be just you and the six women for twenty-eight days and nights on the spaceship. The women are astronauts, doctors, and nurses, and everyone's vital health is monitored and managed for the entire trip. There will be long periods of sleep and then periods of exercise and recreation and an electric piano for you and a variety of instruments for the women to help keep you entertained. You will enjoy the twenty-eight days, and the women will ensure everyone arrives well rested and in peak condition. Some of the women may even have conceived."

Ron smiled when he heard that. Bill communicated to Ron telepathically that there would be a crowd to greet him when he landed and people all over the original world would be watching. They would have medical exams and document everything because it would be history in the making. There would be parties and interest to see this man from another solar system who would be the first ever to physically visit their planet. The men and women from the original planet visited other planets, but no one born on the other planets had stepped onto the original earth. They would take Ron to tour the planet, and he would be somewhat famous, especially with the women who wanted to have his children. They would spend an entire year traveling, and then he could decide where he wanted to spend the next six years and what he wanted to do. They could treat his lung and back and help him cope with his arthritis. Ron heard his inner voice perk up as he thought they hoped he would procreate with as many women as he could during the seven years.

Then he thought, *Hold on. This can't be true. It's like an adolescent dream.*

Ron's inner voice changed to again sound like Bill's, and it said it was true but it would have to be explained later. *There's still potential for this planet to survive, and it's still possible to save it. Procreation of humans from different solar systems should benefit the entire human race, and the outcomes should benefit life on both planets. When one planet dies and the solar system collapses, it causes a reaction through space that potentially kills other solar systems. It may contribute to creating another new solar system with new life, but it also has the potential to kill others, including the original planet. So, it is in human interest and the life of all solar systems for as many as possible to survive. The fact that someone from this planet who shares the DNA of someone from the original planet can procreate indicates they are the same and should be born with the same common knowledge and instincts as both planets. That has the potential to save both planets.*

Ron's own inner voice returned. *The newborns will have DNA from both planets and the instincts and common knowledge of both planets. If everyone on one planet shares the one soul, will a newborn from both planets only be part of one or the other or both? How could you know?*

Then his inner voice sounded like Bill's again and said, *That is what we will find out. A living human being is to the human soul what a single blood cell is to the human body. When blood cells die, new ones are created, and the more new blood cells there are, and the healthier they are, the better the body survives. When a human contributes positively to the human race and procreates and raises offspring to do the same thing, then life and the one soul they all share is better off and survives longer.*

Whether humans or all types of life share the one soul across other solar systems can't be known, but if the newborns who share both DNA are born with the same common knowledge of both planets, it may be indicative. There are some things that will never be known, but we must keep asking questions. Newborns share common instincts and some knowledge at birth, such as knowing it is a newborn human with a mother who has nurtured it from the start and that it will be born into the earth needing to breathe, eat, and grow into an adult. In the womb, the baby has the benefit of being attached to its mother and her brain and will know what it will need to do to survive and grow. If it can be determined a newborn child in either solar system is born with the instincts to survive in the other, that will indicate humans in different solar systems share the same soul. It would not be proof, but it would be significant.

Then Bill said out loud, "There's is just too much to tell you to explain it all right now. A lot of it will come to you when you sleep tonight, and we can talk again later."

Ron said, "Yeah, like why did your planet ever need to speak if you read each other thoughts?"

"We invented language to ask questions. It was only when we didn't know something we needed to learn that we spoke out loud. Now that we travel to other solar systems and need to talk to the local humans, we have all become interested in language even though it is pretty rare to travel that far."

Ron again thought to himself that it was all too crazy and maybe he was dreaming or dead or something.

Bill said, "Yeah, you would be rather tired, too. Your son and all the family will be here soon. They will take you home with them. The boat leaves in two weeks, so you have to decide if you will leave your family for a period of seven years and potentially save your planet and humans in other parts of the universe or stay here on a dying planet."

"When you put it like that," said Ron, "it doesn't seem like much of a choice."

"We're all pretty certain you will come. If you don't, we can take another relative, so you have a choice, but you are our best choice, and we need to know within a few days so we can replace you if we have to."

"Ok, even if I am dead and just dreaming, I may as well go along with it. I'll talk to my family first and get back to you tomorrow."Ron had a new glint in his eyes. "When do I get to meet these six women?"

Bill laughed. "They're around here somewhere. Why don't you wait for your son in the cafeteria? We'll buy you and the family dinner, and the girls can join you."

Chapter Nine
Good Dreams Can Come True

Bill and Ron walked through the halls of Parliament House and into the cafeteria. There were forty or fifty people there eating and about that many empty tables. They sat down at one near the entrance so they would be sure to see and meet Ron's family and the women who would be manning the spacecraft. Ron's family were the first to arrive just minutes later. Ron stood up and walked over to his son, and they hugged. Ron was so elated that his pain was absolutely gone, and he was full of energy.

"I love you guys, he said, "and I knew I would see you again."

He hugged them all together and told them he loved them, and they all returned in kind. Then he hugged each one individually and felt a sensation he interpreted as love as he held them close for several seconds. He had tears in his eyes and a huge smile on his face. He kept looking at their faces while thinking how much he loved them, and he kept hugging and holding hands with his granddaughters. Ron introduced them all to Bill and told them Bill wanted to buy them all dinner.

"We have a lot to talk about," he said, "and there are six other women who will be joining us."

At that moment, the six women walked in the door. Bill said, "Here they are."

Ron stared with his mouth open. They were all beautiful, healthy women, not skinny or fat, but just right. They were all brown-skinned, some darker than others, and they all had long black hair hanging over their shoulders to the middle of their backs. They were all wearing flat shoes, faded blue jeans, and white blouses, and they looked clean, comfortable, and collected. They were different heights, but none of them were very tall, the tallest was probably five and a half feet. Ron looked at the color of their eyes with a huge smile on his face and thought to himself how beautiful they all were. Three of them had brown eyes, and the others had black. Ron thought they all looked like they could be related to one another, or maybe it was just because they were dressed similarly and all had black hair. Two of them had curly hair, and the others had straight. He thought they didn't look like Africans, Asians, Arabs, Anglo-Saxons, or Spanish, but maybe they had mixed parents from one or more of these groups. They were smiling, too, and Bill introduced them all to Ron and his family.

Ron heard their names, but he wasn't paying attention to what was being said. He was looking at the women and thinking it was unbelievable they all wanted to have his child, and his mind was wandering and wondering what it would be like to make love to them. He was still staring with his smiling mouth open when Bill suggested they get in line and get something to eat, and Ron followed them all without saying a word. He was standing at the back of the line, looking at the women from behind and thinking how incredible they looked even from behind.

His oldest granddaughter said, "Grandpa," and he snapped out of it.

"Yes?"

"Did you eat raw fish?"

He said yes, and she squished her nose made a face. He picked her up and hugged her, and the other two granddaughters each hugged one of his legs. His son looked him in the eyes, and Ron said the moment was as good as life could get.

There were plenty of choices of meat or fish, and the servers gave them as much of whatever they wanted. Ron had to consider that his stomach was weak, and he opted for a modest amount of boiled vegetables and a chicken drumstick. When they were all back at the table and had started eating, Bill asked Ron to tell them about being lost at sea. Ron told them the stories of near death and the freakish events that were much like miracles that had saved them. Then they went around the table to let everyone speak about what they had been doing since the first tsunami had hit Sydney. It wasn't planned or stated, but it just happened that way. His granddaughters told him they had been swimming and learning to play their scales on the piano. They were also learning to play drums and were all going to choose another musical instrument in the next year.

His son had been very busy with all kinds of rescues, and he talked about some of them. He said they were lucky in the ACT compared to everywhere else in Australia, but they'd still had a lot of tragedies. So many people across the nation had died that they couldn't even count them all yet, but it was believed to be more than half the population. More than half the landmass was underwater, and Australia now looked like a group of islands. There were the bigger islands of Uluru, the ACT, Blue Mountains, and Victoria's Alps, and the other mountain ranges were groups of islands. Ron's son had been through a lot, too. Apart from the rescues, there had been so many tragedies. He didn't speak of them openly in front of the children, but it was obvious from the things that had happened that he had seen a lot of dead bodies.

Ron's daughter-in-law spoke about the children and what they did when the school was closed. They had been in the garden planting vegetables, and they made the fence around the chicken coop larger and brought home eight more young chickens that would lay eggs soon. They had four laying hens that were still producing, which had provided all the eggs they could eat for the past three years, but they were getting old, and they wanted more eggs to share, so they'd bought more hens. They had given eggs to the neighbors, who were doing it tough from time to time, and after the tsunami, there had been a local food drive, and they had donated four dozen eggs, and the people at the handout had been so happy to get them. That had pleased the girls, and they thought they could help people survive with the additional chooks and give the extra eggs to people who need them. Their mother was pleased they wanted to help people and were willing to help with the chores, so she'd said yes to making the cage bigger and getting more chooks. It wasn't a lot, but it was fresh, healthy food for some, and it would teach the girls to make a contribution to the community and help them feel good about themselves.

The six ladies from the original planet told their stories. They had been on this planet for almost seven years and had seen massive changes in that short time. It had taken them time to adjust to a new way of life with so much violence everywhere. They were horrified by this planet's history of war and violence but had found most people were kind. They had been all together at their shared home in the ACT, watching the news on TV, when the tsunami had hit Sydney, and they described some of the images they'd seen and why they'd decided to volunteer to help at that moment.

"That could have been us on the news," said Ron. "We saw a few drones when we were being washed out of the harbor."

All six girls had volunteered and searched for survivors and then bodies, and after that, they had put bodies into body bags, loaded them onto trucks, and transported them to an outdoor stadium where there were police were trying to identify them. The stories of the number of bodies and piles of body bags and the small number of police to do that kind of work were horrifying. The women had seen a lot of tragedy and had made a huge contribution to the community.

Bill said he had arrived with the women, but unlike them, he was not returning to the original planet. He said he would be doing here what he hoped Ron would be doing on the original planet: prolonging the existence of the human race on both planets. He said that what was good for the human soul was good for the human race and that peace was just one thing that was good for the soul. He said this planet's reaction to climate change had postponed wars between countries for the time being, and that was saving human lives. He said there was still a lot of nuclear fallout that remained airborne and in the sea, so technically, it was the war that would continue to kill humans on this planet for many more years. He said he had great hope and confidence his actions would help stop more wars from starting and eventually stop humans from deliberately killing each other in any manner. He said he was also confident Ron was the person who could best contribute to remedying a threat to the human race on the original planet.

He said Ron carried the DNA in his genes to potentially make an essential contribution to the survival of humans on the original planet and he also had human qualities that women on the original planet found attractive.

Ron wondered why, and his inner voice answered, *Because you're a good guy. You don't lie, cheat, or steal, you respect others, you can be violent but only in self-defense, and you never killed anyone. And the main point would be the shared DNA.* He looked at Bill and heard Bill's voice inside his head say that the DNA and marital status were compulsory, but it was all the other reasons that made him the best choice. The women all clapped after Bill made a toast to the betterment and survival of the human race.

Ron hadn't told his family he was thinking of leaving the planet, but Bill's speech hinted that he was.

Ron's eldest granddaughter said, "Grandpa, are you going somewhere?"

"Not tonight," said Ron, "but maybe pretty soon."

The woman closest to him said, "We all thought you'd made up your mind and were coming with us."

"We're pretty sure I am," said Ron, "but it's not going to be easy to leave these guys for seven years."

"It's for the benefit of all mankind, and we will take good care of you." Then the woman turned to the granddaughter and said, "He can come back in seven years, and all his aches and pains will be better."

His granddaughter nodded without smiling, indicating she understood but would miss her grandpa.

Ron was looking at the woman who was talking, and she stopped and looked into his eyes. At that moment, he knew the answer to the question he had asked himself of how the women who wanted to procreate with him would treat him. The answer from his inner voice said that when it came time to procreate, they would do anything he wanted them to and with gusto. The woman nodded and smiled as if she were reading his mind.

Ron said to his granddaughter, "Yes, I'll be going, but we're going to have a week to be together and have hugs every day, and I'll be back in seven years."

"Let's drink to that," said Bill, and he raised his glass, to which everyone raised theirs and touched it to the glass of the person next to them before having a drink.

Ron thought he might be dead or in a coma and having one heck of a dream, or maybe he was being conned. He was almost certain he wasn't dreaming, but he thought that if he were, it was a good dream, and he would go on with it. He would enjoy the dinner, wake up in the morning, and spend the day with his son and granddaughters, and that would bring plenty of happiness. He could spend the rest of his life at home with his son and his family and be happy, so he didn't need to go anywhere. He could also have a seven-year adventure that sounded too good to be true. He would have to sacrifice seeing his family for seven years, and that was sad, but he would return and see them all again. It could be he was finally being rewarded for trying to live a good life, or it could be a trick, and there was something unseen, and it really was too good to be true.

They finished talking for the evening and agreed to meet and talk more before Ron walked out of the cafeteria with his family. They were walking down the hall when Ron felt giddy and had to stop. His son helped him to remain standing while he regained his wits before they continued.

When they arrived home, Ron could remember holding the hand of one of his granddaughters and being led to the car, and he could recall a vision of headlights on the road at night, but that was the all he could remember from the time he had become giddy. He felt like he had been dreaming or just wasn't totally awake. They all got out of the car and walked into the house, and as soon as they were inside, Ron nearly fainted. His son helped him get to a bed, and his granddaughters hugged and kissed him goodnight and said they would see him in the morning. Ron hoped he would see them in the morning, and he remembered hugging and kissing them all, and he replayed the memory in his head as he fell asleep.

Chapter Ten
The End and the Beginning

Ron woke in darkness. His pain was gone, and then, when he thought about it, he found he couldn't feel anything below his neck. He could only feel his face, which felt fuzzy and numb. He opened his eyes, and it was so dark he was unsure if it was night or if he was blind. He thought about where he was, and his inner voice told him he was at his son's house, in bed, but then he doubted that was true. His inner voice spoke to him quietly and said, *I'm dead.*

If he was dead, how long had he been dead? Maybe he was in a coma and just thought he was dead. He was supposed to be going to another planet to procreate with hundreds of beautiful women to help save the human race all over the universe. If he were to suddenly die before that happened, it would just be cruel, especially after all he'd gone through to survive. How unfair would it be if, just as he was about to enjoy life and at the same time do something good for mankind, he kicked the bucket? Then he thought about how corny that all sounded. Perhaps it was all a dream after all. Maybe it was nature's way of rewarding him for trying so hard to survive and live a good life. He thought that when he kicked the bucket, he would have hoped for a heaven, and that was the reason he'd had the dream that he'd hugged his son and granddaughters and it had felt so real. It was the heaven he'd imagined, and instead of a fearful, sad, painful death, he'd had a wonderful dream.

He wondered about the moment when he'd died. Was it while they were out at sea or even before that, during the Sydney tsunami? It could have been that Rubbish Island and the house on the rock with all those people had been part of the dream, too. He thought that would have been a long dream, but maybe he'd been in a coma or something for a long time and dreaming throughout the ordeal to survive. He recalled the ordeal and thought it was too real to be a dream, and then a vision of him sticking an oar into the mud replayed in his head. He thought it was most likely when they'd been edging the boat along on the mud flats that he'd died. He could never know for certain, and considering his life was over, he couldn't think or any reason he needed to know. How or why he'd died mattered less than how he'd lived and that life had ended. He wondered how he had thoughts if he was dead, and if he was in a coma or just in the last moments of life as he crossed over to death. He thought that was another thing he could never know, but he knew his mortal life had ended.

He recalled his life from his earliest memories, and at the end, when he recalled his dream after death, he fixated on Bill and the other planet. Could that part of the dream have been real or have real meaning? Perhaps Bill was from the original planet and had come to him in a dream.

Ron thought perhaps everything about the original planet was true and could only be learned in a dream. If someone just started talking out loud about other planets and humans on other planets and said they were from the original planet, everyone else would simply call them crazy. Perhaps the people from the original planet could only reach out to people on this planet in a dream after they died. More likely, it was just part of a pleasant dream where he had been able to hug his family and think about procreating with beautiful women as he'd suffered a horrible death, but he could still consider the possibility there was some truth in it. It seemed a gyp to wake up from the dream before he got to fly through space in a spaceship with six beautiful women who wanted to have his baby. He thought the dream was too significant to dismiss, and because he only had his thoughts, he would allow himself to think everything Bill had said was true. There were other planets with human life, and the planet the inhabitants all called the original planet wanted him to come there and procreate with as many women as he could.

His played another re-run of his life in his mind's eye. He felt elated as he recalled the love he'd shared and everyone he'd ever known. He regretted the mistakes in his life and especially the incidents when his actions had harmed or offended others. He thought maybe that was heaven and hell: he was in heaven when he recalled hugging his son or granddaughters, and he was in hell when he remembered when he had insulted or offended someone. It seemed he knew what every person he'd ever known was thinking about him, and when those were good thoughts, he felt good. When those were bad thoughts, he felt sorrow, and that, he thought, could be another interpretation of heaven and hell. He wondered if he would have an eternity to remember his life over and over and was thankful he had never killed anyone and had tried to live a good life even if he'd never achieved anything great. His thoughts returned to the other planets, and it occurred to him he could think of things other than his life, but of course, he could not act on his thoughts unless he were alive, and perhaps that was also heaven and hell.

He thought of the notion that if you can think you exist, you do exist. Then he wondered if existence was possible without actually being alive. He asked himself why he had dreamed about meeting the people from the original planet. They had wanted him to go to their planet and said they would see him later. He thought that could have been significant and perhaps he could live again, or maybe he was alive after all and in some kind of deep trance flying through space in their spaceship and would wake up when they landed on the original planet. He wondered if he could have blacked out before boarding the spaceship and if he would wake from his sleep on the original planet. He hung on to the thought for a moment, and then his inner voice again told him he had died but was still dreaming. He wondered if the dreams had meaning or were random and decided there would be no harm if he kept dreaming.

He remembered in his dream that Bill had said the girls wanted to procreate with him and hoped to conceive during the journey. He hoped that maybe he would dream about that and that one of the girls would wake him at any moment and tell him what was going on. Then he thought all of that was unlikely and he should face up to the fact he had died on the mudflats. He wondered how should he face up to the fact he was dead and what should he be thinking about. He was a little more certain he was dead, but as long as he had thoughts, he could hold hope, and then suddenly, he had a notion his thoughts were keeping him in existence. If he wanted to move from existence to life, perhaps he should keep thinking and not let his thoughts rest. He should just keep thinking and dreaming about anything he could just because he could.

He again recalled all the events of his life and everyone he had ever known over and over. Life had been so hard at times, but it had been wonderful more often. There were bad memories, too, but he would do it all again if he could. If only he could go back and hug his son and granddaughters, but he knew he couldn't. He thought that was why he'd had that dream that had seemed so real, and he replayed all the moments he had ever held them over and over in his mind's eye. He began wishing to be alive again and wondered how it was possible. He was asking himself questions and waiting for answers, but none came. He felt like there were others all around him, but none of them were responding. He wondered if the others were keeping the answers from him or didn't know themselves.

He thought that without life, there was nothing, and he desperately wanted to live again. He was pleading to live again, or at least to be told if and how it was possible. He could sense a dark sky full of stars and infinite space, and he felt it was cold and he could be in space. He was startled when what seemed to be one of the others he had been trying to get answers from coaxed him to face the other direction. When he turned around, he could see the planet earth among millions or billions of other heavenly bodies, and he could sense the other being was advising him to aim for it.

He felt like he was falling and like all the others that had been all around him were watching and becoming further away until he was alone. He wondered if he was falling back to earth and where he had come from or if it could be the original planet he was aiming for.He didn't know where he was, but he knew he was approaching a new life. Maybe that was why he'd had the dream about Bill and had learned all those other things. Perhaps he could be heading towards any one of the other planets Bill had mentioned, and he recalled there were still dinosaurs on some. He felt like himself and wondered if he would have his old body when he arrived and how that could be possible. He thought that the someone or something that had pointed him to earth must have known it was the right place for him to go and that everything would be alright when he landed. Then he recalled he was supposed to aim for it and realized he was free-falling. He tried to look to see ifhe was falling in the right direction but could only see darkness and had no control or notion of where he was going or idea of how he could aim at anything.

He blacked out and then woke up, and he couldn't remember when he had fallen asleep. He was falling fast, and he saw a flash of the planet earth below. He had felt alone before that, but the moment he opened his eyes, he felt others were watching him. There was a loud clap of thunder and a lightning bolt as he approached that seemed to greatly annoy some of the others. They seemed to blame the thunder and lightning on him and asked why he had done that. He thought he didn't know that he had caused the lightning and had no control over what was happening to him. He thought it was natural they were communicating with their thoughts and had received his answer. Then things went absolutely calm, and he blacked out again.

The next instant, he woke, and he knew exactly where he was. He was in a womb. He thought, *Of course, that is where I would have to be to start a new life.* He wondered if he would be the same person in his new life. He had heard somewhere in the past people don't remember previous lives or when they are born, or even when they are very young, and he wondered if he would. He thought the notion about an original planet or going to the original planet had just been a dream, but he wondered if it had significance. Perhaps there were other solar systems and humans just like in his dream. Perhaps there was some kind of truth to it, and he would be born to a mother on the original planet. Maybe he was still the one they'd chosen to bring to the original planet and procreate with as many women possible. Then again, maybe it was just a dream that had made him happy as he died, and that seemed a more likely explanation. He wondered if he would or should remember any of what he was thinking when he was out of the womb. Would it serve any purpose in his new life?

He thought the dream of the women was something he would have dreamt up to fantasize about when he was alive, but perhaps there was a deeper lesson. Maybe the reason the women from the original planet all looked similar, were beautiful, and were descendants of ancestors that had procreated with mates they'd been attracted too because of their different-colored skin was to teach him a lesson about racism. He thought that it couldn't be racism if he admired the race, and he did admire all races. Perhaps the meaning of the dream was to remind him people of all colors on this planet shared the same soul. He hoped he would procreate one day and would consider a mate of any color and teach his children to love people of every color. Perhaps he would forget the dream but remember its meaning and it wasn't something he should ever talk about. If he were to talk about previous lives and other humans living in other solar systems, everyone would think he was crazy and living in some kind of a dream. After all, it really was just a dream, so they might be right.

He was elated to be where he was. He knew all he had to do was eat and grow and that it was up to his mother to take all the responsibility of keeping them alive. The nine months or whatever he was in the womb would be a life of its own. Without any responsibilities, he could enjoy every moment, and in that way, it was better than being outside the womb. He didn't even have to think to eat. The food just went in and out all on its own. He was going to make the most of his time in the womb, and when he was born, he would be the healthiest, happiest baby that could ever be.

He thought about his new mother and wondered what she was like. He knew people were black and white and every shade in between, and he wondered what color she would be. He knew some people had plenty and some had nothing, and he knew there were wars and oppression. He knew there were people who wanted to do good things for the human soul and also people who only cared about themselves and would harm others to get what they wanted. He knew people could kill others for no reason and mistreat each other in horrible ways. He knew there were sharks, wild animals, diseases, and wild weather that could also kill and harm people. People could die of starvation or food poisoning and all kinds of other ways. He knew the world could be a cruel, dangerous place, but it was also possible to find love and happiness, and he decided he absolutely wanted a life. He hoped his mother was safe and her life was good, and he felt she was because he felt healthy. He thought about where he might be born and in what circumstances he would live, and he had fears, but he was elated and looking forward to life.

He grew fast in the first weeks, and even though he had vowed to enjoy every bit of his life inside of the womb, it was already boring. He was happy enough and enjoyed every new thing that happened, but he also seemed to have endless time and kept recalling his previous life even after he told himself to forget it. He wondered if he would remember his previous life after his birth. It could be he would, but because he would be unable to talk until he was around a year old, he might forget everything before he could tell anyone. He wondered if he had been related to his new mother in his previous life. Maybe she was one of his granddaughters or nieces, or maybe humans could begin their new lives with no connection to their past and she was totally unrelated. He recalled the notion of the human race being the one living soul since the beginning of time and believed it. Once he was born, he would still be a part of the one living soul he and everyone else had been, and he wanted to always make positive contributions to keep it in existence. He didn't know what his new life would be, but he believed he would be a good, and maybe a great, man.

Over the next few weeks, he began feeling sick. Things his mother was eating or drinking were causing her to vomit in the mornings, and it was miserable for the baby, too. There were long periods without food, and the baby would get hungry. He had thought all the time in the womb would be a pleasure, but he learned he needed to eat to grow. He started to work hard to eat as much as he could, but he could only eat as much as he was fed, and when he wanted more, he would kick about, trying to make his feelings known. He would eventually wear himself out and go to sleep, and then he would wake up and do it all over again.

There was nothing else he could do, and that gave him plenty of time for his thoughts. He had been holding on to the memories of his past life, but they were less frequent. He wondered if he should try to remember or try to forget. The past was knowledge of some sort, and any knowledge would be a good thing, but maybe he wasn't meant to remember. He thought children always had to learn everything from the beginning, so obviously, they forget their past lives. If a memory of a previous life had to do with DNA, there might be thousands of ancestors capable of giving the child memories of the past. If that happened, the child wouldn't have as much time for their own thoughts, and that was another reason to forget the past life.

He considered his task in the new life was to be a positive contribution to humanity and enjoy it at the same time. What was the point of everyone struggling to survive and trying to make the world a better place if they didn't enjoy life? He would strive to live in the moment, and whenever he had to choose between doing something that was good or bad for the human soul, he would choose what was good. It seemed so simple, and if he did that, there would be no reason or benefit to humanity for him to recall his past life. Still, there he was, recalling his previous life, and it seemed it was impossible to forget. When he was recalling the past life, he wasn't thinking about the present or the future, and that had the potential to be a detriment to his goals. He thought out the pros and cons and decided he should stop trying to recall the previous life and hoped if there was some benefit, he would recall it naturally without effort. He thought perhaps that is what they called it instinct, and he was elated with his decision, and then he noticed he was very hungry and started kicking his mother in the innards.

Weeks passed, and he enjoyed the warmth of the womb and being taken care of. His mother had been sick at times, and there were stretches of time without food, and he knew when he was growing and when he wasn't. He had some very long sleeps or periods of time that he didn't grow, and when he would wake, he wondered what had happened and tried to make up for it by eating faster. He had no control of how he felt or what he ate, and even though he liked being taken care of and was enjoying what he considered to be a life of its own in the womb, he worried about his mother. What was she eating that made her sick? Did she know what she was doing? Maybe there was a good and necessary reason she'd eaten something that had made her sick, or maybe it was an accident. When he felt sick or hungry, he thought he would be hungry and sick forever or would die, and it seemed to be an eternity. Then, eventually, he felt well, and it was so wonderful to feel well that he forgot all about being sick.

More weeks passed, and he learned to tolerate the periods of hunger and pain. He knew that when the hunger pain began, he would eventually be fed, and if he felt sick, he would eventually feel better. Some of the times between feeds were longer than others, but there was always a feed. He was beginning to notice differences in the feed when it dawned on him the food his mother had ingested was the food she was giving to him. There were things he liked and things he wished she wouldn't feed him, but there were a lot more of the good things. He wondered if the things she was eating she liked and if there was some way to let her know what he liked best. He celebrated his newfound discovery and notion with a gyration coupled with a roll around and kicks to the walls of his mother's womb. He could feel her love and wanted her to feel his, and he felt clever because he knew she was feeding him what she was eating. He knew when he was born she would be there, and she would know he loved her.

As time went on, he learned there were other people around his mother. So far, it had only been him and her, but slowly, he became aware of others. He didn't know who the others were, but he knew when there were few or many around his mother. He knew that sometimes, his mother was moving around and things were hectic, and then she would be still, and things went quiet. He slept for long periods, and when he woke up he, was always hungry, and he knew his chances of a quick feed were better when things were still and quiet.

Sometimes, his mother was upset, and that made him sick, and then it dawned on him there were things going on outside of the womb he wasn't part of. He wondered if those others around his mother even knew he was in there. Then it occurred to him he must have a father. He knew there must be a male and female to procreate, but the notion ofhis own father only occurred to him at that moment. He wondered what kind of man his father would be and if he could have brothers or sisters. He had previously thought it was only going to be him and his mother, but then it dawned on him there would be others. He knew those other people around his mother then would also be there when he was born.

When he had first wished for life, he'd imagined it as an adult, and it was only when he was in the womb that he knew he would have to be born and start a brand-new life. He knew a lot about the world but hadn't actually connected all the dots to his mother and him until that moment. He might have brothers and sisters, but those others around his mother could be strangers, too. He remembered there were all kinds of people out there, even some that would harm and kill others. The dangerous people could be out there, and maybe that was why his mother was upset. When he thought about fearful things, there was panic in his stomach, and random thoughts raced through his brain.

It dawned on him the things he had thought about doing would have to wait because he was going to be born a baby and his mother would have to take care of him until he was mature enough to take care of himself. There were all those dangerous things that he knew about, and his mother was just another human amongst them all. He hoped she would be able to cope in the dangerous, cruel world. He trusted she knew how to survive, or he wouldn't be alive.

The next time his mother was upset, he sprang to attention, feeling more urgency than he had ever felt before. There were rapid movements and an awful taste in his mouth that caused him to tighten his stomach. He could hear the volume outside rise, and he began to move his feet and arms back and forth vigorously. He felt fear and panic and wanted to let everyone know he was in there. He could tell there was someone else out there who was upset, too, and that was when he felt some kind of shock. It was like a spark to and from him to a person outside his mother's womb, and he knew that whoever was out there was aware he was in there.

His mother was screaming, and he felt she was in danger. He waited for her to calm down and for things to get quiet, but there were only short pauses and then more noise. When things continued to escalate, he thought his mother was going to die before he was even born. He could sense violence and felt his mother's life was in danger. He desperately wanted her to survive and wanted to get out of the womb to help his mother, but then he remembered he would be born a helpless baby. His mind flashed back to his previous life and recalled his son and granddaughters. He recalled them and his life before he was in the long rowboat lost at sea, and then he recalled how they had all tried so hard to survive. He recalled how, after trying so hard, they had relaxed and slept through storms when there had been nothing else they could do. He thought that was how he could weather the storm he was in, and he relaxed as his mother continued to yell.

Eventually, things calmed down, and he was even more relieved. He thought about what had happened and how he had recalled his previous life after he had decided he would forget about it for the good of his future. He thought about his son and granddaughters and thought there was no way of knowing how much time had passed between the previous life and the present, but he would be born to one of his descendants. He had been trying to forget the previous life, but then it kept popping back into his head, and some of the memories were pleasing. He would be a new individual, and the benefits from his previous life would only be the instincts all living beings receive from their ancestors. He thought it might or might not be a benefit to recall his previous life, but he was enjoying it at that moment.

He recalled his dream of Bill from the original planet, when Bill had said they never had wars or violence there. He knew from the fear he'd felt when his mother had been in danger that he wouldn't be born onto the original planet. He doubted there was an original planet and thought it was only his dream of a better place, but there was nothing wrong with hoping there was such a place. The dreams of other planets and the one shared human soul that survives on each planet from the beginning of that planet's life might not be true, but living a life as if it were would be a good thing because every benefit to the survival of the human soul is also good for the living body.

He was certain this planet was a dangerous place, but the worst thing was that its danger for most humans and every other living thing was because of humans. The wars and murder and greed just did not make sense when he thought about the world he would be born into. He decided the best possible way forward was to strive to be the best human he could be on this world and forget the notion of living on any other planet. He would let his instincts guide him, and in regard to life and humanity, he would always choose right from wrong. It would be his task in life to procreate and keep the human soul alive on this planet. He would procreate and then raise and love his children and teach them to know what was right or wrong and encourage them to always do what was right. He would be born and live his life on this world, whichever world it was, and he would do everything he could to make this world a better place for the one living soul all humans share, and then he would die again.

The End

www.ingramcontent.com/pod-product-compliance
Lightning Source LLC
Chambersburg PA
CBHW060354220326
41598CB00023B/2914